计算机技术开发与应用丛书

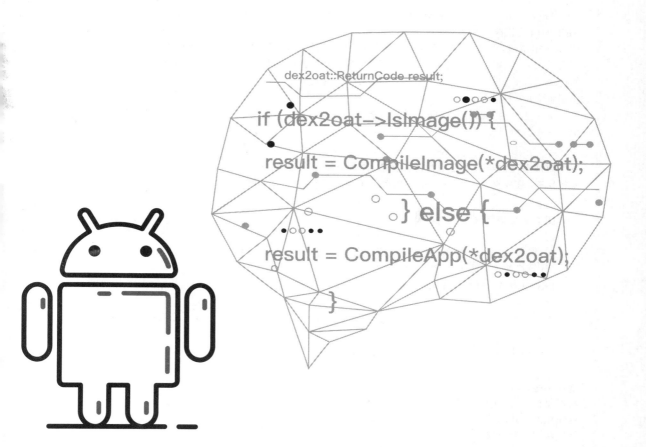

Android Runtime
源码解析

史宁宁 ◎ 编著

清华大学出版社

北京

内 容 简 介

Android Runtime(ART)作为 Android 系统的核心组件,是 Android 应用运行及其效率的基础,了解 ART 的组成和执行流程,有利于 Android 应用开发者高效地开发 Android 应用程序。同时,ART 作为一种 Java 虚拟机,也是广大虚拟机开发者学习和改进的目标,了解 ART 有助于虚拟机的开发和调优。

本书基于 Android 10.0.0_r39 源码,分析 ART 基础、ART 的 compiler、ART 的启动与运行、ART 中的垃圾回收等。这些分析旨在为读者搭建 ART 的基本框架,带领读者一起了解 ART 的基本情况,为进一步了解 ART 打下基础。

本书适合想了解 ART 的 App 开发者、Android 系统优化开发者及虚拟机开发者阅读,也适合高校学生学习虚拟机时使用。

本书封面贴有清华大学出版社防伪标签,无标签者不得销售。
版权所有,侵权必究。举报: 010-62782989,beiqinquan@tup.tsinghua.edu.cn。

图书在版编目(CIP)数据

Android Runtime 源码解析/史宁宁编著. —北京: 清华大学出版社,2022.4
(计算机技术开发与应用丛书)
ISBN 978-7-302-60084-8

Ⅰ. ①A… Ⅱ. ①史… Ⅲ. ①移动终端-应用程序-程序设计 Ⅳ. ①TN929.53

中国版本图书馆 CIP 数据核字(2022)第 023993 号

责任编辑: 赵佳霓
封面设计: 吴　刚
责任校对: 时翠兰
责任印制: 朱雨萌

出版发行: 清华大学出版社
网　　址: http://www.tup.com.cn,http://www.wqbook.com
地　　址: 北京清华大学学研大厦 A 座　　邮　编: 100084
社 总 机: 010-83470000　　邮　购: 010-62786544
投稿与读者服务: 010-62776969,c-service@tup.tsinghua.edu.cn
质量反馈: 010-62772015,zhiliang@tup.tsinghua.edu.cn
课件下载: http://www.tup.com.cn,010-83470236

印 装 者: 北京鑫海金澳胶印有限公司
经　　销: 全国新华书店
开　　本: 186mm×240mm　　印　张: 18　　字　数: 408 千字
版　　次: 2022 年 6 月第 1 版　　印　次: 2022 年 6 月第 1 次印刷
印　　数: 1~2000
定　　价: 79.00 元

产品编号: 090618-01

序
FOREWORD

在浩瀚的 Android 源码中，Android Runtime(ART)作为一种 JVM，是其中最为人知晓但又最为陌生的软件模块。无数 App 运行在 JVM 上，但又有多少人研读过它的代码呢？本人不才，在 2019 年撰写过一本有关 Android 7.0 中 JVM ART 源代码分析的书。不过，Android 发展很快，ART 的代码也在更新，它是一个有生命力的软件模块，所以我一直有一个希望，就是有一些志同道合的朋友能持续跟踪 ART 并继续研究和分析它。毕竟，追求百战百胜之前，先做到知己知彼是必不可少的。

当我看完《Android Runtime 源码解析》一书，顿时感觉自己的小小希望终于得以实现。本书作者史宁宁挑选了 Android 10.0 中 ART 的一些最为核心的部分，如 Java 字节码编译、ART GC 原理和实现等，进行了更为详尽的介绍。虽然书的厚度中规中矩，但涉及知识的难度绝对不小。

最后，我还要特别推介的是作者所在的团队——中国科学院软件研究所 PLCT 实验室。我在工作中和 PLCT 实验室诸位同僚有过多次深入接触。目前在人们更多关注应用领域创新的时候，我对该实验室"致力于成为编译技术领域的开源领导者，推进开源工具链及运行时系统等软件基础设施的技术革新，具备主导开发和维护重要基础设施的技术及管理能力；与此同时，努力成为编译领域培养尖端人才的黄埔军校，推动先进编译技术在国内的普及和发展"的目标甚为支持。在基础技术领域里吸引和培养更多年轻才俊确实是到了刻不容缓的时候了。我看到了 PLCT 实验室付出的诸多努力，也希望借助本书，能吸引更多的读者参与进来。

邓凡平

"深入理解 Android"系列丛书的主要作者

前言
PREFACE

Android 已经成为当前应用最广泛的手机操作系统之一。无论是 Android 应用开发者，还是 Android 生态系统开发者，甚至是操作系统设计者，都有熟悉 Android Runtime（ART）的需求。

本书从一个编译器开发者的视角，带领读者在 ART 的世界里遨游，和大家一起了解 ART 的各部分及其主要流程。

在本书的编写过程中，力图将 ART 的整体架构梳理清楚，包括在介绍其中的模块时，也是将架构介绍清楚作为第一目标，尽量避免太多细节内容。代码总在不断地更新，但是模块架构和整体架构不会变更得那么频繁。掌握好架构，既可以快速地了解 ART 的整体情况，也有利于自己根据需要对某些模块进行深入研究。同时，为了让读者能熟悉最新的代码，本书选用了 Android 10.0.0_r39 的代码，读者可以采用该版本的代码对照本书进行学习。

本书的内容分为 4 部分。第 1 部分包括第 1 章，主要对 ART 的基础进行介绍；第 2 部分包括第 2~4 章，主要介绍 ART 中的 compiler 部分，包括 dex2oat 工具，属于编译时阶段；第 3 部分包括第 5、6 章，主要介绍 ART 的启动及运行，属于运行时阶段；第 4 部分包括第 7 章，主要介绍 ART 中的垃圾回收部分。读者阅读本书的时候，可以按照顺序从头至尾阅读，也可以根据需要直接选择其中某一部分内容进行阅读，并不会影响对相关内容的理解。

各章的内容分别如下：第 1 章从虚拟机基础、ART 发展历史、ART 核心架构和源码目录结构等方面，介绍 ART 基础；第 2 章介绍 dex2oat 工具的入口、driver 及 DexToDexCompiler 等；第 3 章对 OptimizingCompiler 中的 JNI 处理、Compile 过程进行分析，并对 Compile 过程中的主要环节——展开进行介绍；第 4 章对 OptimizingCompiler 中与硬件平台无关和与硬件平台相关的优化进行介绍，并选取与硬件平台无关优化中的一些典型优化进行深入分析；第 5 章分析 ART 在启动时的几个主要流程；第 6 章对 ART 在执行时的主要流程进行分析；第 7 章对 ART GC 的整体架构、种类及具体实现进行分析。

相对于 ART 的庞大内容，本书所介绍的内容并不能完全覆盖所有内容，读者在阅读本书时应注意结合源码，从源码中获取更多的细节内容。

由于作者水平有限，书中难免存在疏漏，敬请读者批评指正。

史宁宁
2022 年 5 月

本书配套代码

目 录
CONTENTS

第 1 章 ART 基础 ··· 1
    1.1 虚拟机基础 ··· 1
    1.2 ART 发展历史 ·· 2
    1.3 ART 核心架构 ·· 3
    1.4 源码目录介绍 ··· 4
    1.5 小结 ·· 4

第 2 章 dex2oat 工具介绍 ··· 5
    2.1 dex2oat 入口代码分析 ·· 5
    2.2 dex2oat 的 driver 分析 ·· 7
    2.3 dex2oat driver 的编译函数 ··· 18
    2.4 DexToDexCompiler 分析 ··· 19
    2.5 小结 ·· 27

第 3 章 OptimizingCompiler 介绍 ··· 29
    3.1 OptimizingCompiler 类的 JNI 处理 ·· 29
    3.2 OptimizingCompiler：：Compile() ·· 35
    3.3 HGraph 的构建 ·· 42
        3.3.1 构建基于基本代码块的 CFG ·· 44
        3.3.2 构建支配树 ·· 45
        3.3.3 构建 SSA ··· 49
    3.4 优化 ·· 51
    3.5 寄存器分配 ·· 59
        3.5.1 PrepareForRegisterAllocation ·· 60
        3.5.2 SsaLivenessAnalysis ·· 61
        3.5.3 RegisterAllocator ··· 62

3.6 代码生成 ··· 64
3.7 OptimizingCompiler 总结 ·· 69
3.8 小结 ··· 70

第 4 章 OptimizingCompiler 优化算法分析 ·· 71

4.1 优化算法框架 ·· 71
4.2 常量折叠 ·· 79
4.3 指令简化 ·· 90
4.4 死代码优化 ··· 99
4.5 循环体优化 ·· 104
4.6 指令下沉 ·· 109
4.7 硬件平台相关优化 pass 及其实现 ·· 115
4.8 小结 ··· 120

第 5 章 ART 启动分析 ··· 122

5.1 ART 启动中的虚拟机启动一 ··· 122
5.2 ART 启动中的虚拟机启动二 ··· 133
5.3 ART 启动中的 JIT 编译器的创建 ·· 138
5.4 ART 启动中的 Thread 处理 ··· 142
5.5 ART 启动中的运行时本地方法初始化 ·· 153
5.6 ART 启动中的其他本地方法的注册 ··· 166
5.7 Zygote 进程 ·· 175
　　5.7.1 System Server 进程 ··· 179
　　5.7.2 应用进程 ·· 185
5.8 小结 ··· 190

第 6 章 ART 的执行 ·· 191

6.1 ART 运行基本流程 ·· 191
6.2 Zygote 进程调用应用程序 ··· 192
　　6.2.1 Zygote.forkAndSpecialize ·· 197
　　6.2.2 ZygoteConnection.handleChildProc ·· 199
6.3 类的查找与定义 ··· 205
6.4 方法的加载和链接 ·· 220
6.5 方法的执行 ·· 222
6.6 小结 ··· 230

第 7 章　ART GC 实现 ··· 231
7.1　GC 的基本内容 ·· 231
7.2　ART GC 回收方案介绍 ·· 233
7.3　ART GC 回收器的实现 ·· 234
7.3.1　回收器的类型 ·· 234
7.3.2　不同类型回收器的实现 ·· 238
7.4　ART GC 的分配器实现 ·· 246
7.4.1　分配器与空间类 ··· 246
7.4.2　分配器与回收器 ··· 256
7.5　ART GC 的使用流程 ·· 257
7.5.1　分配器的使用 ·· 258
7.5.2　回收器的使用 ·· 263
7.6　小结 ·· 275

参考文献 ·· 276

后记 ··· 277

第1章 ART 基础

随着 Android 系统的广泛应用，Android Runtime(ART)作为 Android 系统的核心，引起了广大 Android 开发人员及系统研究人员的重视，越来越多的人开始深入地研究 ART 的内部实现，本章将简要介绍与 ART 相关的基础内容。

1.1 虚拟机基础

虚拟机作为一个近年来逐渐发展壮大的技术领域，在模拟各种硬件平台乃至为程序语言提供解释执行平台等方面，应用得越来越广泛。本部分内容将就虚拟机的基础知识进行介绍，为读者展示一个基本的虚拟机知识体系。

虚拟机(Virtual Machine,VM)通常是指由软件所实现的硬件架构。虚拟机分为进程虚拟机(Process VM)和系统虚拟机(System VM)。进程虚拟机实现了一个 ABI(Application Binary Interface)，这个 ABI 是由用户层的 ISA(Instruction Set Architecture)和操作系统的系统调用接口组成的。系统虚拟机实现了一整套的硬件，包括硬件用户(Hardware User)和系统 ISA。

进程虚拟机的典型代表是高级语言虚拟机，这类虚拟机通常是为了实现高级语言的可移植性，进而采用虚拟机支持高级语言或者高级语言的 IR 进行解释执行。高级语言虚拟机的典型代表就是 Java 虚拟机(JVM)。在 JVM 的体系中，Java 程序被 javac 翻译成 Bytecode 格式后运行在 JVM 之上。这里的 JVM 作为一个高级语言虚拟机，它为 Bytecode 的运行提供了用户层的 JVM ISA，同时 JVM 作为一个应用程序是以操作系统的一个独立进程在运行，如图 1.1 所示。

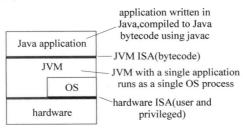

图 1.1 JVM 示意图

(图源：UCB CS294：Virtual Machines and Managed Runtimes-Slide 1 Intro，overview，classification)

系统虚拟机的典型代表是 VirtualBox。VirtualBox 可以在已有的主机系统（Host System）之上再安装一个客户机系统（Guest System），而这个客户机系统依然是一个完整的操作系统，应用程序在客户机系统上运行和在主机系统上运行没有什么差别，并且，VirtualBox 上还可以建立多个客户机系统，这些客户机系统可以和主机系统完全不同。VirtualBox 为客户机系统提供了一个完整的硬件平台，这个硬件平台是包含硬件用户和系统 ISA 的，如图 1.2 所示。

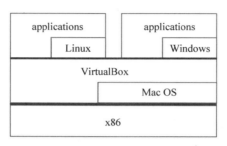

图 1.2　VirtualBox 示意图

（图源：UCB CS294：Virtual Machines and Managed Runtimes-Slide 1 Intro，overview，classification）

JVM 和 ART（Android Runtime）从广义上都属于 Java 的虚拟机，但是二者还有不少区别。JVM 的输入是 Java 的 Bytecode 字节码，ART 的输入是 dex 字节码。dex 字节码最初是为 ART 的前身 Dalvik 虚拟机设计的一种压缩格式，适用于内存和处理器速度有限的系统；二者在结构上也有很大区别。JVM 依然是解释执行和即时编译模式，而 ART 已经开始侧重于 AOT 编译，加快应用程序的启动时间并减少运行时内存占用，所以二者在实现上有很大的不同，在进行相关学习时应注意区分。

1.2　ART 发展历史

Android 作为目前最流行的移动设备操作系统，广泛应用于手机、平板电脑、智能设备等产品之上。随着 Android 的广泛应用，Android 应用的数量也呈指数级增长，这两者的发展都对 Android 性能不断提出更高的要求。Android 性能的核心部分是其虚拟机，这包括早期的 Dalvik 和现在的 ART。

Dalvik 虚拟机是 Android 系统的核心组成部分之一，它由 Dan Bornstein 编写，名字来源于 Dan Bornstein 祖先曾经居住过的一个小渔村（Dalvik）。Dalvik 虚拟机是一种基于寄存器的虚拟机，它接受专门为其设计的.dex 格式的输入。.dex 格式是通过 dx 工具从 Java class 转化而来的，多个类会包含在一个 dex 文件中。Dalvik 虚拟机早期只能进行解释执行，为了提升其执行速度，从 Android 2.2 版本开始，Dalvik 虚拟机也支持了即时编译技术（Just In Time，JIT）。

随着 Android 系统的发展，提升 Android 系统上应用的运行速度成为一个重要的问题。ART 随之被提出，它作为 Android 系统的运行时环境，其实是 Android 的一种新虚拟机，并且是作为 Dalvik 的替代品出现的。ART 最早出现在 Android 4.4 中,当时还只是一项测试功能,到 Android 5.0 及其以后的版本中,ART 已经作为正式组件替代了 Dalvik 虚拟机。

ART 作为新一代的虚拟机,它的核心采用了预编译技术(Ahead of Time,AOT)。预编译技术会在应用程序安装过程中将字节码直接编译为机器码,在运行时可以直接调用运行,它比之前 Dalvik 所使用的即时编译技术在性能上有很大的提升。ART 以预编译技术为主,但是其依然保留了即时编译的部分。ART 作为新一代虚拟机,与 Dalvik 虚拟机相比,其也具有一定的缺点。应用程序在安装时,ART 会比 Dalvik 需要更多的安装时间和应用存储空间。

1.3 ART 核心架构

AOSP 项目的资料中有关于 ART 的核心架构图,具体如图 1.3 所示。

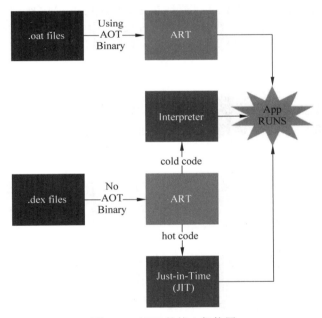

图 1.3　ART 的核心架构图

(图源:https://source.android.com/devices/tech/dalvik/jit-compiler)

从图 1.3 中可以看到,ART 的执行方式分为 AOT、JIT 和 Interpreter 3 种。其中,AOT 是提前编译,它将输入的 dex 格式编译为 oat 格式进行输出,这也是 dex2oat 工具所在做的事情,而这部分的核心代码实现并不在 art/dex2oat 中,而是位于 art/compiler/ 中,后续会根据具体内容进行具体介绍。JIT 是即时编译,将解释执行过程中的热点代码通过编译器编译之后,直接执行编译之后的版本去替代未编译的代码,以此来提高效率。JIT 这种方式,在早期的 Android 中是提高 App 运行效率的主要方式。Interpreter 是解释执行,这种模式是从最早期的 Android 一直延续下来的执行模式,也是 Java 程序主要采用的模式,当然 ART 中的 Interpreter 和通常所使用的 Java Interpreter 还是有所不同的。

1.4 源码目录介绍

ART 的源码是由一个独立的库管理的,它的网址为 https://android.googlesource.com/platform/art/,可以单独拉去阅读。在发布的各个版本的 Android 源码中,ART 位于源码目录之下的 art 目录。Android 发布版本的时候,会选择所需要的 ART 版本和系统一起发布。

在 art 目录下有 20 多个目录,其中常用的有 dex2oat、compiler、Runtime 等。dex2oat 目录主要存放的内容是 dex2oat 工具的接口的相关代码;compiler 目录存放的是进行 AOT 编译的相关代码;Runtime 存放的是运行时的相关代码。本书所进行分析的代码和实现,也主要位于这 3 个目录。其他重要的目录是在 ART 使用过程中所用到的库和工具代码目录。

1.5 小结

本章从虚拟机的基础入手,介绍了 ART 的发展历史,并对 ART 的核心架构及主要代码目录进行了简单介绍,期望为读者展示简要的 ART 全景。

第 2 章 dex2oat 工具介绍

dex2oat 工具是 ART 的主要工具,负责对 dex 进行编译处理。在 dex2oat 中,一个 Java 方法根据其模式,可以将 dex 字节码编译成为 oat 格式,也可以对 dex 字节码进行进一步的优化。本章将对 dex2oat 工具的主要执行流程的代码实现进行详细介绍。

2.1 dex2oat 入口代码分析

分析一个工具的代码实现,首先要对该工具代码的入口函数进行分析,这就好比是一个宫殿的大门,只有找到了大门,才能进一步登堂入室。dex2oat 工具的入口函数是位于 art/dex2oat/dex2oat.cc 中的 main()函数,调用 dex2oat 的工具都是从这个函数开始执行的。

在这个 main()函数中,它调用了 dex2oat.cc 中的 art::Dex2oat()函数,而 Dex2oat()函数中进行编译工作的函数主要是 dex2oat.cc 中的 CompileImage()函数和 CompileApp()函数,它们分别根据所对应的不同格式进行编译处理,代码如下:

```
//第 2 章/dex2oat.cc
  dex2oat::ReturnCode result;
  if (dex2oat->IsImage()) {
    result = CompileImage(*dex2oat);
  } else {
    result = CompileApp(*dex2oat);
  }
```

这里根据 dex2oat->IsImage()的值来选择调用函数的路径。其中,CompileImage()函数先调用了 LoadClassProfileDescriptors()函数,加载了类配置的描述,然后调用了 dex2oat 的 Compile()函数,代码如下:

```
//第 2 章/dex2oat.cc
static dex2oat::ReturnCode CompileImage(Dex2Oat& dex2oat) {
  dex2oat.LoadClassProfileDescriptors();
  jobject class_loader = dex2oat.Compile();
  …
```

CompileApp 中也调用了 dex2oat::Compile() 函数,代码如下:

```
//第2章/dex2oat.cc
static dex2oat::ReturnCode CompileApp(Dex2Oat& dex2oat) {
    jobject class_loader = dex2oat.Compile();
    …
```

由于上述两个路径都调用了 dex2oat.cc 中的 dex2oat::Compile() 函数,就等于两个路径最终都会走到 dex2oat::Compile() 函数。dex2oat::Compile() 函数的核心代码是创建了一个新的 CompilerDriver 对象 driver_,这个 driver_ 是通过 reset 的形式将一个新的 CompilerDriver 对象设置到 driver_中,之后调用了 CompileDexFiles() 函数,代码如下:

```
//第2章/dex2oat.cc
//Set up and create the compiler driver and then invoke it to compile
//all the dex files
jobject Compile() {
    ClassLinker * const class_linker =
        Runtime::Current()->GetClassLinker();
…
    driver_.reset(new CompilerDriver(compiler_options_.get(),
                                compiler_kind_,
                                thread_count_,
                                swap_fd_));
…
    return CompileDexFiles(dex_files);
}
```

上面调用的 CompileDexFiles() 函数实现于 dex2oat.cc 中,在它的具体实现中,它通过 driver_调用了 PreCompile() 函数和 CompileAll() 函数等进行编译 dex 格式所需要的核心函数,代码如下:

```
//第2章/dex2oat.cc
//Create the class loader, use it to compile, and return
jobject CompileDexFiles(const std::vector<const DexFile *>& dex_files) {
    ClassLinker * const class_linker =
        Runtime::Current()->GetClassLinker();
…
    driver_->InitializeThreadPools();
    driver_->PreCompile(class_loader,
                        dex_files,
                        timings_,
                        &compiler_options_->image_classes_,
                        verification_results_.get());
```

```
      callbacks_->SetVerificationResults(nullptr);
      compiler_options_->verification_results_ =
          verification_results_.get();
      driver_->CompileAll(class_loader, dex_files, timings_);
      driver_->FreeThreadPools();
      return class_loader;
    }
```

这里需要特别说明，driver_属于CompilerDriver类型的指针，它的定义也在dex2oat.cc中，代码如下：

```
    std::unique_ptr<CompilerDriver> driver_;
```

所以，后续对于代码的跟踪，就要转入CompilerDriver类的实现中。

本部分介绍了dex2oat工具的入口函数，并按照入口函数逐层根据调用关系向下进行代码跟踪，最终跟踪到了driver_的成员函数的调用，并且可以从中看出这里调用的driver_的成员函数都是进行dex字节码编译的函数。同时，也可以看到，本部分分析的代码都位于dex2oat.cc中，算是宏观意义上的入口函数，接下来的代码执行就要进入CompilerDriver的实现，笔者将在下一部分继续进行分析和介绍。

2.2 dex2oat 的 driver 分析

上文从dex2oat工具的入口函数main()开始进行分析，根据调用流程分析到对CompilerDriver类成员函数的调用，本部分将对CompilerDriver类成员函数进行分析。CompilerDriver类可以被视为dex2oat工具的driver部分。

CompilerDriver类的定义和实现位于art/dex2oat/driver目录的compiler_driver.h和compiler_driver.cc中。上文提到的CompilerDriver类的成员函数PreCompile()和CompileAll()的实现都位于compiler_driver.cc中。CompileAll()函数中的核心是对Compile()函数的调用，代码如下：

```
//第2章/compiler_driver.cc
void CompilerDriver::CompileAll(
    jobject class_loader,
    const std::vector<const DexFile*>& dex_files,
    TimingLogger* timings) {
  DCHECK(!Runtime::Current()->IsStarted());

  …
  //Compile
  //1) Compile all classes and methods enabled for compilation. May fall
  //back to dex-to-dex compilation
```

```
    if (GetCompilerOptions().IsAnyCompilationEnabled()) {
      Compile(class_loader, dex_files, timings);
    }
    if (GetCompilerOptions().GetDumpStats()) {
      stats_->Dump();
    }
}
```

Compile()函数也是CompilerDriver类的成员函数,它的实现位于compiler_driver.cc中。它的主要功能是编译所有可以编译的类和方法,并且这里的编译不仅是dex-to-oat的编译,还有可能会返回dex-to-dex的编译,所以也可以看到在调用Compile()函数前,需要有编译选项。Compile()函数的实现,在核心部分两次调用了CompileDexFile()函数,两者不同的地方在于传入的最后一个参数不同,这个参数是回调函数,两次分别传入了CompileMethodQuick()函数和CompileMethodDex2Dex()函数,将dex格式编译为oat格式或者对dex进行优化,这里也就对应了前文提到的对dex文件的两种处理方式。Compile()函数的代码如下:

```
//第2章/compiler_driver.cc
void CompilerDriver::Compile(
    jobject class_loader,
    const std::vector<const DexFile*>& dex_files,
    TimingLogger* timings) {
  if (kDebugProfileGuidedCompilation) {
    const ProfileCompilationInfo* profile_compilation_info =
        GetCompilerOptions().GetProfileCompilationInfo();
    LOG(INFO) << "[ProfileGuidedCompilation] " <<
        ((profile_compilation_info == nullptr)
          ? "null"
          : profile_compilation_info->DumpInfo(dex_files));
  }

  dex_to_dex_compiler_.ClearState();
  for (const DexFile* dex_file : dex_files) {
    CHECK(dex_file != nullptr);
    CompileDexFile(this,
                   class_loader,
                   *dex_file,
                   dex_files,
                   parallel_thread_pool_.get(),
                   parallel_thread_count_,
                   timings,
                   "Compile Dex File Quick",
                   CompileMethodQuick);
```

```
    const ArenaPool* const arena_pool =
        Runtime::Current()->GetArenaPool();
    const size_t arena_alloc = arena_pool->GetBytesAllocated();
    max_arena_alloc_ = std::max(arena_alloc, max_arena_alloc_);
    Runtime::Current()->ReclaimArenaPoolMemory();
  }

  if (dex_to_dex_compiler_.NumCodeItemsToQuicken(Thread::Current()) > 0u) {
    //TODO: Not visit all of the dex files, its probably rare that only
    //one would have quickened methods though
    for (const DexFile* dex_file : dex_files) {
      CompileDexFile(this,
                     class_loader,
                     *dex_file,
                     dex_files,
                     parallel_thread_pool_.get(),
                     parallel_thread_count_,
                     timings,
                     "Compile Dex File Dex2Dex",
                     CompileMethodDex2Dex);
    }
    dex_to_dex_compiler_.ClearState();
  }

  VLOG(compiler) << "Compile: " << GetMemoryUsageString(false);
}
```

上文两次调用了 CompileDexFile() 函数,并且通过参数的不同,为其设置了不同的回调函数,进而完成了 dex 的两种编译途径的执行。需要注意的是,CompileDexFile() 函数并不是 CompilerDriver 类的成员函数,虽然它的实现也位于 compiler_driver.cc 中,但是它是函数模板,其中类型参数用于最后一个参数,用来接收回调函数。CompileDexFile() 函数在具体实现过程中,通过 context.ForAllLambda() 函数(位于同文件中)发起线程执行编译处理,最终将其操作转入了回调函数 compile_fn() 中,代码如下:

```
//第 2 章/compiler_driver.cc
template <typename CompileFn>
static void CompileDexFile(
    CompilerDriver* driver,
    jobject class_loader,
    const DexFile& dex_file,
    const std::vector<const DexFile*>& dex_files,
    ThreadPool* thread_pool,
    size_t thread_count,
    TimingLogger* timings,
```

```cpp
                const char* timing_name,
                CompileFn compile_fn) {
  TimingLogger::ScopedTiming t(timing_name, timings);
  ParallelCompilationManager context(
      Runtime::Current()->GetClassLinker(),
      class_loader,
      driver,
      &dex_file,
      dex_files,
      thread_pool);

  auto compile = [&context, &compile_fn](size_t class_def_index) {
    const DexFile& dex_file = *context.GetDexFile();
    SCOPED_TRACE << "compile " << dex_file.GetLocation()
                 << "@" << class_def_index;
    ClassLinker* class_linker = context.GetClassLinker();
    jobject jclass_loader = context.GetClassLoader();
    ClassReference ref(&dex_file, class_def_index);
    const dex::ClassDef& class_def =
        dex_file.GetClassDef(class_def_index);
    ClassAccessor accessor(dex_file, class_def_index);
    CompilerDriver* const driver = context.GetCompiler();
    //Skip compiling classes with generic verifier failures since they
    //will still fail at Runtime
    if (driver->GetCompilerOptions().GetVerificationResults()->
        IsClassRejected(ref)) {
      return;
    }
    //Use a scoped object access to perform to the quick SkipClass
    //check
    ScopedObjectAccess soa(Thread::Current());
    StackHandleScope<3> hs(soa.Self());
    Handle<mirror::ClassLoader> class_loader(
        hs.NewHandle(soa.Decode<mirror::ClassLoader>(jclass_loader)));
    Handle<mirror::Class> klass(
        hs.NewHandle(class_linker->FindClass(soa.Self(),
          accessor.GetDescriptor(), class_loader)));
    Handle<mirror::DexCache> dex_cache;
    if (klass == nullptr) {
      soa.Self()->AssertPendingException();
      soa.Self()->ClearException();
      dex_cache = hs.NewHandle(class_linker->FindDexCache(soa.Self(), dex_file));
    } else if (SkipClass(jclass_loader, dex_file, klass.Get())) {
      return;
    } else if (&klass->GetDexFile() != &dex_file) {
      //Skip a duplicate class (as the resolved class is from another, earlier dex file)
```

```cpp
      return;
      //Do not update state.
    } else {
      dex_cache = hs.NewHandle(klass->GetDexCache());
    }

    //Avoid suspension if there are no methods to compile
    if (accessor.NumDirectMethods() + accessor.NumVirtualMethods() ==
        0) {
      return;
    }

    //Go to native so that we don't block GC during compilation
    ScopedThreadSuspension sts(soa.Self(), kNative);

    //Can we run DEX-to-DEX compiler on this class
    optimizer::DexToDexCompiler::CompilationLevel
        dex_to_dex_compilation_level = GetDexToDexCompilationLevel(
            soa.Self(), *driver, jclass_loader, dex_file, class_def);

    //Compile direct and virtual methods
    int64_t previous_method_idx = -1;
    for (const ClassAccessor::Method& method : accessor.GetMethods()) {
      const uint32_t method_idx = method.GetIndex();
      if (method_idx == previous_method_idx) {
        //smali can create dex files with two encoded_methods sharing the same method_idx
        //http://code.google.com/p/smali/issues/detail?id=119
        continue;
      }
      previous_method_idx = method_idx;
      compile_fn(soa.Self(),
                 driver,
                 method.GetCodeItem(),
                 method.GetAccessFlags(),
                 method.GetInvokeType(class_def.access_flags_),
                 class_def_index,
                 method_idx,
                 class_loader,
                 dex_file,
                 dex_to_dex_compilation_level,
                 dex_cache);
    }
  };
  context.ForAllLambda(0, dex_file.NumClassDefs(), compile,
                       thread_count);
}
```

根据上文可知，CompileDexFile 所接收的回调函数为 CompileMethodQuick()和 CompileMethodDex2Dex()。其中，CompileMethodQuick()函数虽然位于compiler_driver.cc 中，但是它也不是 CompilerDriver 类的成员函数。CompileMethodQuick()函数主要通过在内部构建 quick_fn()函数之后，调用 CompileMethodHarness()函数完成操作，CompileMethodHarness()函数调用了 quick_fn()回调函数。其中，需要特别指出的是，quick_fn()函数在构建 compiled_method 的时候分为两种：driver—>GetCompiler()—>JniCompile()和 driver—>GetCompiler()—>Compile()。

CompileMethodQuick()函数的代码如下：

```
//第2章/compiler_driver.cc
static void CompileMethodQuick(
    Thread* self, CompilerDriver* driver,
    const dex::CodeItem* code_item,
    uint32_t access_flags, InvokeType invoke_type,
    uint16_t class_def_idx, uint32_t method_idx,
    Handle<mirror::ClassLoader> class_loader,
    const DexFile& dex_file,
    optimizer::DexToDexCompiler::CompilationLevel
        dex_to_dex_compilation_level,
    Handle<mirror::DexCache> dex_cache) {
  auto quick_fn = [](
      Thread* self,
      CompilerDriver* driver,
      const dex::CodeItem* code_item,
      uint32_t access_flags,
      InvokeType invoke_type,
      uint16_t class_def_idx,
      uint32_t method_idx,
      Handle<mirror::ClassLoader> class_loader,
      const DexFile& dex_file,
      optimizer::DexToDexCompiler::CompilationLevel
          dex_to_dex_compilation_level,
      Handle<mirror::DexCache> dex_cache) {
    DCHECK(driver != nullptr);
    CompiledMethod* compiled_method = nullptr;
    MethodReference method_ref(&dex_file, method_idx);

    if ((access_flags & kAccNative) != 0) {
  //Are we extracting only and have support for generic JNI down calls
      if (!driver->GetCompilerOptions().IsJniCompilationEnabled() &&
          InstructionSetHasGenericJniStub(driver->GetCompilerOptions()
              .GetInstructionSet())) {
        //Leaving this empty will trigger the generic JNI version
      } else {
```

```cpp
        //Query any JNI optimization annotations such as @FastNative or @CriticalNative
        access_flags |= annotations::GetNativeMethodAnnotationAccessFlags(
            dex_file, dex_file.GetClassDef(class_def_idx), method_idx);

        compiled_method = driver->GetCompiler()->JniCompile(
            access_flags, method_idx, dex_file, dex_cache);
        CHECK(compiled_method != nullptr);
      }
    } else if ((access_flags & kAccAbstract) != 0) {
      //Abstract methods don't have code
    } else {
      const VerificationResults* results =
          driver->GetCompilerOptions().GetVerificationResults();
      DCHECK(results != nullptr);
      const VerifiedMethod* verified_method =
          results->GetVerifiedMethod(method_ref);
      bool compile =
          //Basic checks, e.g., not <clinit>
          results->IsCandidateForCompilation(method_ref, access_flags) &&
          //Did not fail to create VerifiedMethod metadata
          verified_method != nullptr &&
          //Do not have failures that should punt to the interpreter
          !verified_method->HasRuntimeThrow() &&
          (verified_method->GetEncounteredVerificationFailures() &
              (verifier::VERIFY_ERROR_FORCE_INTERPRETER | verifier::VERIFY_ERROR_LOCKING))
              == 0 &&
              //Is eligible for compilation by methods-to-compile filter
              driver->ShouldCompileBasedOnProfile(method_ref);

      if (compile) {
        //NOTE: if compiler declines to compile this method, it will
        //return null
        compiled_method = driver->GetCompiler()->Compile(code_item,
                                                         access_flags,
                                                         invoke_type,
                                                         class_def_idx,
                                                         method_idx,
                                                         class_loader,
                                                         dex_file,
                                                         dex_cache);
      ProfileMethodsCheck check_type =
          driver->GetCompilerOptions().CheckProfiledMethodsCompiled();
      if (UNLIKELY(check_type != ProfileMethodsCheck::kNone)) {
        bool violation = driver->ShouldCompileBasedOnProfile(method_ref)
            &&(compiled_method == nullptr);
        if (violation) {
```

```
            std::ostringstream oss;
            oss << "Failed to compile "
                << method_ref.dex_file->PrettyMethod(method_ref.index)
                << "[" << method_ref.dex_file->GetLocation() << "]"
                << " as expected by profile";
            switch (check_type) {
              case ProfileMethodsCheck::kNone:
                break;
              case ProfileMethodsCheck::kLog:
                LOG(ERROR) << oss.str();
                break;
              case ProfileMethodsCheck::kAbort:
                LOG(FATAL_WITHOUT_ABORT) << oss.str();
                _exit(1);
            }
          }
        }
      }
      if (compiled_method == nullptr &&
          dex_to_dex_compilation_level !=
              optimizer::DexToDexCompiler::CompilationLevel
                  ::kDontDexToDexCompile) {
        DCHECK(!Runtime::Current()->UseJitCompilation());
        driver->GetDexToDexCompiler().MarkForCompilation(self,
            method_ref);
      }
    }
    return compiled_method;
  };
  CompileMethodHarness(self,
                       driver,
                       code_item,
                       access_flags,
                       invoke_type,
                       class_def_idx,
                       method_idx,
                       class_loader,
                       dex_file,
                       dex_to_dex_compilation_level,
                       dex_cache,
                       quick_fn);
}
```

上面介绍了 CompileMethodQuick() 函数, 和它一起被调用的还有一个 CompileMethodDex2Dex() 函数, CompileMethodDex2Dex() 函数的情况和 CompileMethodQuick()

函数的情况类似。CompileMethodDex2Dex()函数也位于 compiler_driver.cc 中,并且不是 CompilerDriver 类的成员函数,它也是通过新建了一个 dex_2_dex_fn()函数,最终通过 CompileMethodHarness()函数回调执行 dex_2_dex_fn()函数。CompileMethodDex2Dex() 函数的代码如下:

```
//第 2 章/compiler_driver.cc
static void CompileMethodDex2Dex(
    Thread* self,
    CompilerDriver* driver,
    const dex::CodeItem* code_item,
    uint32_t access_flags,
    InvokeType invoke_type,
    uint16_t class_def_idx,
    uint32_t method_idx,
    Handle<mirror::ClassLoader> class_loader,
    const DexFile& dex_file,
    optimizer::DexToDexCompiler::CompilationLevel
        dex_to_dex_compilation_level,
    Handle<mirror::DexCache> dex_cache) {
  auto dex_2_dex_fn = [](Thread* self ATTRIBUTE_UNUSED,
    CompilerDriver* driver,
    const dex::CodeItem* code_item,
    uint32_t access_flags,
    InvokeType invoke_type,
    uint16_t class_def_idx,
    uint32_t method_idx,
    Handle<mirror::ClassLoader> class_loader,
    const DexFile& dex_file,
    optimizer::DexToDexCompiler::CompilationLevel
        dex_to_dex_compilation_level,
    Handle<mirror::DexCache> dex_cache ATTRIBUTE_UNUSED) ->
        CompiledMethod* {
  DCHECK(driver != nullptr);
  MethodReference method_ref(&dex_file, method_idx);

  optimizer::DexToDexCompiler* const compiler =
      &driver->GetDexToDexCompiler();

  if (compiler->ShouldCompileMethod(method_ref)) {
    const VerificationResults* results =
        driver->GetCompilerOptions().GetVerificationResults();
    DCHECK(results != nullptr);
    const VerifiedMethod* verified_method =
        results->GetVerifiedMethod(method_ref);
    //Do not optimize if a VerifiedMethod is missing. SafeCast elision,
```

```
            //for example, relies on it
            return compiler -> CompileMethod(
                code_item,
                access_flags,
                invoke_type,
                class_def_idx,
                method_idx,
                class_loader,
                dex_file,
                (verified_method != nullptr)
                ? dex_to_dex_compilation_level
                    : optimizer::DexToDexCompiler::CompilationLevel
                        ::kDontDexToDexCompile);
        }
        return nullptr;
    };
    CompileMethodHarness(self,
                         driver,
                         code_item,
                         access_flags,
                         invoke_type,
                         class_def_idx,
                         method_idx,
                         class_loader,
                         dex_file,
                         dex_to_dex_compilation_level,
                         dex_cache,
                         dex_2_dex_fn);
}
```

所以,compiler_driver.cc 中的 CompileDexFile() 函数会调用 CompileMethodQuick() 函数和 CompileMethodDex2Dex() 函数,而这两个函数其实最终都被执行到 CompileMethodHarness() 函数中,而位于 compiler_driver.cc 中的 CompileMethodHarness() 函数,其实就是根据所接收的回调函数 quick_fn() 或 dex_2_dex_fn() 执行,然后返回编译完的方法,在代码中就是 quick_fn compiled_method() 函数和 dex_2_dex_fn() 函数中的 compiler -> CompileMethod() 函数继续执行。CompileMethodHarness() 函数的代码如下:

```
//第2章/compiler_driver.cc
template < typename CompileFn >
static void CompileMethodHarness(
    Thread* self,
    CompilerDriver* driver,
    const dex::CodeItem* code_item,
    uint32_t access_flags,
```

```cpp
      InvokeType invoke_type,
      uint16_t class_def_idx,
      uint32_t method_idx,
      Handle<mirror::ClassLoader> class_loader,
      const DexFile& dex_file,
      optimizer::DexToDexCompiler::CompilationLevel
          dex_to_dex_compilation_level,
      Handle<mirror::DexCache> dex_cache,
      CompileFn compile_fn) {
  DCHECK(driver != nullptr);
  CompiledMethod* compiled_method;
  uint64_t start_ns = kTimeCompileMethod ? NanoTime() : 0;
  MethodReference method_ref(&dex_file, method_idx);

  compiled_method = compile_fn(self, driver, code_item, access_flags,
                               invoke_type, class_def_idx, method_idx,
                               class_loader, dex_file,
                               dex_to_dex_compilation_level,
                               dex_cache);

  if (kTimeCompileMethod) {
    uint64_t duration_ns = NanoTime() - start_ns;
    if (duration_ns > MsToNs(driver->GetCompiler()->
                                    GetMaximumCompilationTimeBeforeWarning())) {
      LOG(WARNING) << "Compilation of "
                   << dex_file.PrettyMethod(method_idx)
                   << " took " << PrettyDuration(duration_ns);
    }
  }

  if (compiled_method != nullptr) {
    driver->AddCompiledMethod(method_ref, compiled_method);
  }

  if (self->IsExceptionPending()) {
    ScopedObjectAccess soa(self);
    LOG(FATAL) << "Unexpected exception compiling: "
               << dex_file.PrettyMethod(method_idx) << "\n"
               << self->GetException()->Dump();
  }
}
```

在实际执行中，CompileMethodHarness()函数中要编译 Compile Method 而调用的函数其实一共有 3 个，它们分别是：driver->GetCompiler()->JniCompile()、driver->GetCompiler()->Compile()和 compiler->CompileMethod()。这 3 个函数分别处理不

同情况的method,对其进行编译,这3个函数的具体实现将在2.3节中进行详细介绍。

本节主要针对CompilerDriver类的成员函数,以及compiler_driver.cc中的几个非CompilerDriver类但又与driver相关的函数进行了分析,总体而言这些代码都属于compiler_driver.cc,即属于driver部分的代码。本节所分析的内容,涵盖了从与dex2oat的入口函数相关的处理之后的driver环节,对于driver环节所涉及的3个编译method的函数,在后续内容中将进一步介绍。

2.3　dex2oat driver 的编译函数

dex2oat的driver部分涉及对method进行编译的3个函数,这3个函数分别为driver->GetCompiler()->JniCompile()、driver->GetCompiler()->Compile()和compiler->CompileMethod()。前文已经分别介绍了从dex2oat的入口main()函数到driver的这3个函数的调用关系,本部分将对这3个函数的实现进行分析。

总体看来,这3个函数都是由CompilerDriver先通过GetCompiler()函数或者GetDexToDexCompiler()函数获取对应的Compiler或者DexToDexCompiler,最终调用Compiler或者DexToDexCompiler的编译函数。GetCompiler()函数的定义位于compiler_driver.h中,代码如下:

```
//第2章/compiler_driver.h
  Compiler* GetCompiler() const {
    return compiler_.get();
  }
```

其中compiler_是一个Compiler类型的指针,代码如下:

```
std::unique_ptr<Compiler> compiler_;
```

GetDexToDexCompiler()函数的定义位于compiler_driver.h中,代码如下:

```
//第2章/compiler_driver.h
  optimizer::DexToDexCompiler& GetDexToDexCompiler() {
    return dex_to_dex_compiler_;
  }
```

dex_to_dex_compiler_是DexToDexCompiler类型,代码如下:

```
//Compiler for dex to dex (quickening)
optimizer::DexToDexCompiler dex_to_dex_compiler_;
```

同时,compiler_初始化的时候,是在compiler_driver.cc中CompilerDriver的构造函数

中，代码如下：

```
compiler_.reset(Compiler::Create( * compiler_options,
                    &compiled_method_storage_, compiler_kind));
```

所以，3个编译函数可以概括为一个 Compiler 或其子类的 JniCompile() 函数和 Compile() 函数，一个 DexToDexCompiler 或其子类的 CompileMethod() 函数。

Compiler 的定义和实现位于 art/compiler/目录中的 compiler.h 和 compiler.cc 中，它是一个基类。Compiler 在整个 Android ART 源码中只有一个子类：OptimizingCompiler，定义和实现位于 art/compiler/optimizing_compiler.cc 中，代码如下：

```
class OptimizingCompiler final : public Compiler {
```

所以，这里的 JniCompile() 函数和 Compile() 函数都是 OptimizingCompiler 的成员函数，而 DexToDexCompiler 的定义和实现位于 art/dex2oat/dex/目录的 dex_to_dex_compiler.h 和 dex_to_dex_compiler.cc 中，并没有子类，所以 CompileMethod() 函数指的是 DexToDexCompiler 的 CompileMethod() 函数。

至此，dex2oat 中有关编译的 3 个函数已经找到最终的指向，同时每个函数都代表着一类编译。JniCompile() 函数将 Jni 方法编译到 oat，Compile() 函数将正常的 dex 编译到 oat，CompileMethod() 函数将 dex 编译到 dex。这正好对应了关于 dex2oat 的 3 种编译配置。

2.4 DexToDexCompiler 分析

上文提到了 3 个对 method 进行编译的函数，并且对这 3 个函数的具体实现进行了跟踪。其中，DexToDexCompiler 的 CompileMethod() 函数位于 dex2oat 目录，所以本节将对其具体实现进行介绍。其余两个函数的实现，因为其不在 dex2oat 目录，将在后续章节中介绍。

DexToDexCompiler 类的定义和实现位于 art/dex2oat/dex/目录的 dex_to_dex_compiler.h 和 dex_to_dex_compiler.cc 中。DexToDexCompiler 类的 CompileMethod() 函数主要用于实现从 dex 到 dex 的编译，这一过程也可以看作对 dex 格式文件的优化。DexToDexCompiler 类的 CompileMethod() 函数的具体实现位于 dex_to_dex_compiler.cc 中，它的核心是调用了关键函数 CompilationState::Compile() 和 CompiledMethod::SwapAllocCompiledMethod() 实现了其功能。CompileMethod() 函数的代码如下：

```
//第2章/dex_to_dex_compiler.cc
CompiledMethod * DexToDexCompiler::CompileMethod(
    const dex::CodeItem * code_item,
```

```cpp
    uint32_t access_flags,
    InvokeType invoke_type ATTRIBUTE_UNUSED,
    uint16_t class_def_idx,
    int32_t method_idx,
    Handle<mirror::ClassLoader> class_loader,
    const DexFile& dex_file,
    CompilationLevel compilation_level) {
  if (compilation_level == CompilationLevel::kDontDexToDexCompile) {
    return nullptr;
  }

  ScopedObjectAccess soa(Thread::Current());
  StackHandleScope<1> hs(soa.Self());
  ClassLinker* const class_linker = Runtime::Current()->GetClassLinker();
  art::DexCompilationUnit unit(
      class_loader,
      class_linker,
      dex_file,
      code_item,
      class_def_idx,
      method_idx,
      access_flags,
      driver_->GetCompilerOptions().GetVerifiedMethod(&dex_file, method_idx), hs.
NewHandle(class_linker->
            FindDexCache(soa.Self(), dex_file)));

  std::vector<uint8_t> quicken_data;
  //If the code item is shared with multiple different method ids, make sure that we quicken
  //only once and verify that all the dequicken
  //maps match
  if (UNLIKELY(shared_code_items_.find(code_item) !=
      shared_code_items_.end())) {
    //Avoid quickening the shared code items for now because the
    //existing conflict detection logic does not currently handle cases where the code item
    //is quickened in one place but
    //compiled in another
    static constexpr bool kAvoidQuickeningSharedCodeItems = true;
    if (kAvoidQuickeningSharedCodeItems) {
      return nullptr;
    }
    //For shared code items, use a lock to prevent races
    MutexLock mu(soa.Self(), lock_);
    auto existing = shared_code_item_quicken_info_.find(code_item);
    QuickenState* existing_data = nullptr;
    std::vector<uint8_t>* existing_quicken_data = nullptr;
    if (existing != shared_code_item_quicken_info_.end()) {
```

```cpp
    existing_data = &existing->second;
    if (existing_data->conflict_) {
      return nullptr;
    }
    existing_quicken_data = &existing_data->quicken_data_;
  }
  bool optimized_return_void;
  {
    CompilationState state(this, unit, compilation_level,
        existing_quicken_data);
    quicken_data = state.Compile();
    optimized_return_void = state.optimized_return_void_;
  }

  //Already quickened, check that the data matches what was previously seen
  MethodReference method_ref(&dex_file, method_idx);
  if (existing_data != nullptr) {
    if ( * existing_quicken_data != quicken_data ||
        existing_data->optimized_return_void_ !=
            optimized_return_void) {
      VLOG(compiler) << "Quicken data mismatch, for method "
                    << dex_file.PrettyMethod(method_idx);
      //Mark the method as a conflict to never attempt to quicken it in the future
        existing_data->conflict_ = true;
    }
    existing_data->methods_.push_back(method_ref);
  } else {
    QuickenState new_state;
    new_state.methods_.push_back(method_ref);
    new_state.quicken_data_ = quicken_data;
    new_state.optimized_return_void_ = optimized_return_void;
    bool inserted = shared_code_item_quicken_info_.emplace(code_item,
        new_state).second;
    CHECK(inserted) << "Failed to insert "
                    << dex_file.PrettyMethod(method_idx);
  }

  //Easy sanity check is to check that the existing stuff matches by re-quickening using
  //the newly produced quicken data
  //Note that this needs to be behind the lock for this case since we may unquicken in
  //another thread
  if (kIsDebugBuild) {
    CompilationState state2(this, unit, compilation_level,
        &quicken_data);
    std::vector<uint8_t> new_data = state2.Compile();
    CHECK(new_data == quicken_data) <<
```

```
          "Mismatch producing new quicken data";
    }
  } else {
    CompilationState state(this, unit, compilation_level,
        /* quicken_data */ nullptr);
    quicken_data = state.Compile();

    //Easy sanity check is to check that the existing stuff matches by re-quickening using
    //the newly produced quicken data
    if (kIsDebugBuild) {
      CompilationState state2(this, unit, compilation_level,
          &quicken_data);
      std::vector<uint8_t> new_data = state2.Compile();
      CHECK(new_data == quicken_data) <<
          "Mismatch producing new quicken data";
    }
  }

  if (quicken_data.empty()) {
    return nullptr;
  }

  //Create a `CompiledMethod`, with the quickened information in the
  //vmap table
  InstructionSet instruction_set =
      driver_->GetCompilerOptions().GetInstructionSet();
  if (instruction_set == InstructionSet::kThumb2) {
    //Don't use the thumb2 instruction set to avoid the one off code
    //delta
    instruction_set = InstructionSet::kArm;
  }
  CompiledMethod* ret = CompiledMethod::SwapAllocCompiledMethod(
      driver_->GetCompiledMethodStorage(),
      instruction_set,
      ArrayRef<const uint8_t>(),                    //no code
      ArrayRef<const uint8_t>(quicken_data),  //vmap_table
      ArrayRef<const uint8_t>(),                    //cfi data
      ArrayRef<const linker::LinkerPatch>());
  DCHECK(ret != nullptr);
  return ret;
}
```

CompilationState::Compile()的实现也位于dex_to_dex_compiler.cc中,它是dex2dex的核心函数,通过switch-case的形式逐条处理dex指令,为后续准备编译完成的方法提供了核心数据,代码如下:

```cpp
//第2章/dex_to_dex_compiler.cc
std::vector<uint8_t> DexToDexCompiler::CompilationState::Compile() {
  DCHECK_EQ(compilation_level_, CompilationLevel::kOptimize);
  const CodeItemDataAccessor& instructions = unit_.GetCodeItemAccessor();
  for (DexInstructionIterator it = instructions.begin();
       it != instructions.end(); ++it) {
    const uint32_t dex_pc = it.DexPc();
    Instruction* inst = const_cast<Instruction*>(&it.Inst());

    if (!already_quickened_) {
      DCHECK(!inst->IsQuickened());
    }

    switch (inst->Opcode()) {
      case Instruction::RETURN_VOID:
        CompileReturnVoid(inst, dex_pc);
        break;

      case Instruction::CHECK_CAST:
        inst = CompileCheckCast(inst, dex_pc);
        if (inst->Opcode() == Instruction::NOP) {
          //We turned the CHECK_CAST into two NOPs, avoid visiting the
          //second NOP twice since this would add 2 quickening info
          //entries
          ++it;
        }
        break;

      case Instruction::IGET:
      case Instruction::IGET_QUICK:
        CompileInstanceFieldAccess(inst, dex_pc, Instruction::IGET_QUICK, false);
        break;

      case Instruction::IGET_WIDE:
      case Instruction::IGET_WIDE_QUICK:
        CompileInstanceFieldAccess(inst, dex_pc, Instruction::IGET_WIDE_QUICK, false);
        break;

      case Instruction::IGET_OBJECT:
      case Instruction::IGET_OBJECT_QUICK:
        CompileInstanceFieldAccess(inst, dex_pc, Instruction::IGET_OBJECT_QUICK, false);
        break;

      case Instruction::IGET_BOOLEAN:
      case Instruction::IGET_BOOLEAN_QUICK:
        CompileInstanceFieldAccess(inst, dex_pc, Instruction::IGET_BOOLEAN_QUICK, false);
```

```cpp
      break;

    case Instruction::IGET_BYTE:
    case Instruction::IGET_BYTE_QUICK:
      CompileInstanceFieldAccess(inst, dex_pc, Instruction::IGET_BYTE_QUICK, false);
      break;

    case Instruction::IGET_CHAR:
    case Instruction::IGET_CHAR_QUICK:
      CompileInstanceFieldAccess(inst, dex_pc, Instruction::IGET_CHAR_QUICK, false);
      break;

    case Instruction::IGET_SHORT:
    case Instruction::IGET_SHORT_QUICK:
      CompileInstanceFieldAccess(inst, dex_pc, Instruction::IGET_SHORT_QUICK, false);
      break;

    case Instruction::IPUT:
    case Instruction::IPUT_QUICK:
      CompileInstanceFieldAccess(inst, dex_pc, Instruction::IPUT_QUICK, true);
      break;

    case Instruction::IPUT_BOOLEAN:
    case Instruction::IPUT_BOOLEAN_QUICK:
      CompileInstanceFieldAccess(inst, dex_pc, Instruction::IPUT_BOOLEAN_QUICK, true);
      break;

    case Instruction::IPUT_BYTE:
    case Instruction::IPUT_BYTE_QUICK:
      CompileInstanceFieldAccess(inst, dex_pc, Instruction::IPUT_BYTE_QUICK, true);
      break;

    case Instruction::IPUT_CHAR:
    case Instruction::IPUT_CHAR_QUICK:
      CompileInstanceFieldAccess(inst, dex_pc, Instruction::IPUT_CHAR_QUICK, true);
      break;

    case Instruction::IPUT_SHORT:
    case Instruction::IPUT_SHORT_QUICK:
      CompileInstanceFieldAccess(inst, dex_pc, Instruction::IPUT_SHORT_QUICK, true);
      break;

    case Instruction::IPUT_WIDE:
    case Instruction::IPUT_WIDE_QUICK:
      CompileInstanceFieldAccess(inst, dex_pc, Instruction::IPUT_WIDE_QUICK, true);
      break;
```

```cpp
    case Instruction::IPUT_OBJECT:
    case Instruction::IPUT_OBJECT_QUICK:
      CompileInstanceFieldAccess(inst, dex_pc, Instruction::IPUT_OBJECT_QUICK, true);
      break;

    case Instruction::INVOKE_VIRTUAL:
    case Instruction::INVOKE_VIRTUAL_QUICK:
      CompileInvokeVirtual(inst, dex_pc,
                           Instruction::INVOKE_VIRTUAL_QUICK, false);
      break;

    case Instruction::INVOKE_VIRTUAL_RANGE:
    case Instruction::INVOKE_VIRTUAL_RANGE_QUICK:
      CompileInvokeVirtual(inst, dex_pc,
                           Instruction::INVOKE_VIRTUAL_RANGE_QUICK,
                           true);
      break;

    case Instruction::NOP:
      if (already_quickened_) {
        const uint16_t reference_index = NextIndex();
        quickened_info_.push_back(QuickenedInfo(dex_pc,
                                  reference_index));
        if (reference_index == DexFile::kDexNoIndex16) {
          //This means it was a normal nop and not a check-cast
          break;
        }
        const uint16_t type_index = NextIndex();
        if (driver_.IsSafeCast(&unit_, dex_pc)) {
          quickened_info_.push_back(QuickenedInfo(dex_pc, type_index));
        }
        ++it;
      } else {
        //We need to differentiate between check cast inserted NOP and normal NOP, put an
        //invalid index in the map for normal nops.
        //This should be rare in real code
        quickened_info_.push_back(QuickenedInfo(dex_pc, DexFile::kDexNoIndex16));
      }
      break;

    default:
      //Nothing to do
      break;
  }
}

if (already_quickened_) {
```

```
      DCHECK_EQ(quicken_index_, existing_quicken_info_.NumIndices());
  }

  //Even if there are no indicies, generate an empty quicken info so
  //that we know the method was quickened

  std::vector<uint8_t> quicken_data;
  if (kIsDebugBuild) {
    //Double check that the counts line up with the size of the quicken info
    size_t quicken_count = 0;
    for (const DexInstructionPcPair& pair : instructions) {
      if (QuickenInfoTable::NeedsIndexForInstruction(&pair.Inst())) {
        ++quicken_count;
      }
    }
    CHECK_EQ(quicken_count, GetQuickenedInfo().size());
  }

  QuickenInfoTable::Builder builder(&quicken_data, GetQuickenedInfo().size());
  //Length is encoded by the constructor
  for (const CompilationState::QuickenedInfo& info : GetQuickenedInfo())
  {
    //Dex pc is not serialized, only used for checking the instructions
    //Since we access the array based on the index of the quickened
    //instruction, the indexes must line up perfectly
    //The reader side uses the NeedsIndexForInstruction function too
    const Instruction& inst = instructions.InstructionAt(info.dex_pc);
    CHECK(QuickenInfoTable::NeedsIndexForInstruction(&inst)) << inst.Opcode();
    builder.AddIndex(info.dex_member_index);
  }
  DCHECK(!quicken_data.empty());
  return quicken_data;
}
```

CompilationState::Compile()函数逐条处理完 dex 指令之后，生成了 quicken_data。这个 quicken_data 是下一步构建编译完成的方法的核心数据。CompilationState::Compile()函数执行完后，紧接着执行了 CompiledMethod::SwapAllocCompiledMethod()函数，它是 CompiledMethod 类的成员函数。CompiledMethod 类的定义和实现位于 art/compiler 的 compiled_method.h、compiled_method-inl.h 和 compiled_method.cc 中。CompiledMethod 类用来表示已经编译完成的方法，继承于同个文件中的 CompiledCode 类。SwapAllocCompiledMethod()函数的具体实现位于 compiled_method.cc 中，代码如下：

```
//第 2 章/compiled_method.cc
CompiledMethod* CompiledMethod::SwapAllocCompiledMethod(
    CompiledMethodStorage* storage,
```

```
    InstructionSet instruction_set,
    const ArrayRef<const uint8_t>& quick_code,
    const ArrayRef<const uint8_t>& vmap_table,
    const ArrayRef<const uint8_t>& cfi_info,
    const ArrayRef<const linker::LinkerPatch>& patches) {
  SwapAllocator<CompiledMethod> alloc(storage->GetSwapSpaceAllocator());
  CompiledMethod* ret = alloc.allocate(1);
  alloc.construct(ret,
                  storage,
                  instruction_set,
                  quick_code,
                  vmap_table,
                  cfi_info, patches);
  return ret;
}
```

SwapAllocCompiledMethod()函数最终调用了 SwapAllocator 的 construct()函数。这个函数的实现位于 art/compiler/utils/swap_space.h 中，代码如下：

```
//第2章/swap_space.h
  void construct(pointer p, const_reference val) {
    new (static_cast<void*>(p)) value_type(val);
  }
  template<class U, class... Args>
  void construct(U* p, Args&&... args) {
    ::new (static_cast<void*>(p)) U(std::forward<Args>(args)...);
  }
```

SwapAllocCompiledMethod()函数通过调用 SwapAllocator 的 construct()函数，完成了一个新构建的 CompiledMethod，这里存储了一个编译完的方法，这样就完成了一种从 dex 到 dex 的编译方法。至此，dex2oat 中的 dex2dex 的执行流程已经梳理完毕。

2.5 小结

dex2oat 作为 Android ART 的重要工具，其对 dex 有 3 种处理方式。本部分内容从 dex2oat 的入口函数 mian()函数开始介绍，一直介绍到具体 3 种方式所采用的 compiler 的编译函数。这 3 个编译函数可以直接获取编译完成后的方法，即 CompiledMethod 对象。

本部分对于 dex2dex 的 DexToDexCompiler::CompileMethod()函数的具体实现进行了深入跟踪，而对于其他两个编译函数没有进行深入介绍。这是因为 DexToDexCompiler 的实现处于 art/dex2oat 目录之内，而其他两个编译函数所属的 OptimizingCompiler 实现则位于 art/compiler/optimizing 目录之下，虽然 dex2oat 对它有调用关系，但是还是单独介绍比较好，所以有关 OptimizingCompiler 的另外两个函数，会在介绍 OptimizingCompiler 内容时再专门介绍，在本部分了解到 OptimizingCompiler::JniCompile()函数和 OptimizingCompiler::Compile()函数层面即可。

同时，本部分内容所分析的源码也覆盖了 dex2oat 下的大多数目录，包括 art/dex2oat、art/dex2oat/driver 和 art/dex2oat/dex 目录，同时还涉及 art/compiler 目录的内容，读者可以根据已经分析的内容了解各个目录所包含的具体内容，这将有利于读者自行阅读未分析的代码。

第 3 章 OptimizingCompiler 介绍

OptimizingCompiler 是 ART 中非常重要的一部分，art/compiler 中源码的核心就是 OptimizingCompiler。OptimizingCompiler 从理解上可以分为宏观的 OptimizingCompiler 和微观的 OptimizingCompiler。宏观的 OptimizingCompiler 可以被视为一个完整的编译器，将输入 dex 格式文件编译为 oat 格式，是 dex2oat 工具的核心；微观的 OptimizingCompiler 是一个具体的类，是 Compiler 的子类。本部分内容将对宏观的 OptimizingCompiler 所涉及的主要内容进行介绍，它的源码位于 art/compiler/optimizing 目录。

需要专门指出的是，除非专门说明为微观的 OptimizingCompiler 或者 OptimizingCompiler 类，默认情况下所提的 OptimizingCompiler 指的是宏观的 OptimizingCompiler。

3.1 OptimizingCompiler 类的 JNI 处理

前文介绍 dex2oat 时已经知道对于 dex 有 3 种处理方式，并且分析了其中的 dex2dex 处理方式，对于其中的 Jni 处理对应的 OptimizingCompiler::JniCompile()函数和 dex 到机器码所对应的 OptimizingCompiler::Compile()函数并未介绍。本部分将对 OptimizingCompiler::JniCompile()函数及其所涉及的流程进行介绍。

dex2oat 的 driver 部分通过 driver—>GetCompiler()—>JniCompile()这种形式，最终调用了 OptimizingCompiler::JniCompile()函数。OptimizingCompiler::JniCompile()函数的实现位于 art/compiler/optimizing/optimizing_compiler.cc 中。

OptimizingCompiler::JniCompile()函数负责编译 JNI 方法，在编译过程中，会根据编译选项中是否应编译为 Boot Image，而有不同的执行路径。在确认编译为 Boot Image，并且方法是 Intrinsic 的情况下，通过 TryCompileIntrinsic()函数生成一个 CodeGenerator 指针，然后通过 Emit()函数生成编译过的方法。在编译后不为 Boot Image 的情况下，通过 ArtQuickJniCompileMethod()函数和 CompiledMethod::SwapAllocCompiledMethod()函数生成编译过的方法，代码如下：

```
//第 3 章/optimizing_compiler.cc
CompiledMethod* OptimizingCompiler::JniCompile(uint32_t access_flags,
```

```cpp
    uint32_t method_idx, const DexFile& dex_file,
    Handle<mirror::DexCache> dex_cache) const {
Runtime* Runtime = Runtime::Current();
ArenaAllocator allocator(Runtime->GetArenaPool());
ArenaStack arena_stack(Runtime->GetArenaPool());

const CompilerOptions& compiler_options = GetCompilerOptions();
if (compiler_options.IsBootImage()) {
  ScopedObjectAccess soa(Thread::Current());
  ArtMethod* method = Runtime->GetClassLinker()->LookupResolvedMethod(
      method_idx, dex_cache.Get(), /* class_loader = */ nullptr);
  if (method != nullptr && UNLIKELY(method->IsIntrinsic())) {
    VariableSizedHandleScope handles(soa.Self());
    ScopedNullHandle<mirror::ClassLoader> class_loader;
      Handle<mirror::Class> compiling_class = handles.NewHandle(method->GetDeclaringClass());
    DexCompilationUnit dex_compilation_unit(
        class_loader,
        Runtime->GetClassLinker(),
        dex_file,
        /* code_item = */ nullptr,
        /* class_def_idx = */ DexFile::kDexNoIndex16,
        method_idx,
        access_flags,
        /* verified_method = */ nullptr,
        dex_cache,
        compiling_class);
    CodeVectorAllocator code_allocator(&allocator);
    //Go to native so that we don't block GC during compilation
    ScopedThreadSuspension sts(soa.Self(), kNative);
    std::unique_ptr<CodeGenerator> codegen(
        TryCompileIntrinsic(&allocator,
            &arena_stack,
            &code_allocator,
            dex_compilation_unit,
            method,
            &handles));
    if (codegen != nullptr) {
      CompiledMethod* compiled_method = Emit(&allocator,
          &code_allocator, codegen.get(), /* item = */ nullptr);
      compiled_method->MarkAsIntrinsic();
      return compiled_method;
    }
  }
}
```

```
JniCompiledMethod jni_compiled_method = ArtQuickJniCompileMethod(
    compiler_options, access_flags, method_idx, dex_file);
MaybeRecordStat(compilation_stats_.get(), MethodCompilationStat::kCompiledNativeStub);

//Will hold the stack map
ScopedArenaAllocator stack_map_allocator(&arena_stack);
ScopedArenaVector<uint8_t> stack_map =
    CreateJniStackMap(&stack_map_allocator, jni_compiled_method);
return CompiledMethod::SwapAllocCompiledMethod(
    GetCompiledMethodStorage(),
    jni_compiled_method.GetInstructionSet(),
    jni_compiled_method.GetCode(),
    ArrayRef<const uint8_t>(stack_map),
    jni_compiled_method.GetCfi(),
    /* patches = */ ArrayRef<const linker::LinkerPatch>());
}
```

上文提到了对 TryCompileIntrinsic( )函数的调用，TryCompileIntrinsic( )函数的内部实现可以分为构建 HGraph、运行优化（RunOptimizations）、运行目标优化（RunArchOptimizations）、寄存器分配和代码生成（codegen—>Compile）等几个主要环节，之后返回了一个 CodeGenerator 指针。TryCompileIntrinsic( )函数的这几个环节，每个环节都是编译的一个重要过程，这个部分将会在后续专门讨论，在此不再展开，代码如下：

```
//第 3 章/optimizing_compiler.cc
CodeGenerator* OptimizingCompiler::TryCompileIntrinsic(
    ArenaAllocator* allocator,
    ArenaStack* arena_stack,
    CodeVectorAllocator* code_allocator,
    const DexCompilationUnit& dex_compilation_unit,
    ArtMethod* method,
    VariableSizedHandleScope* handles) const {
MaybeRecordStat(compilation_stats_.get(), MethodCompilationStat::kAttemptIntrinsicCompilation);
const CompilerOptions& compiler_options = GetCompilerOptions();
InstructionSet instruction_set = compiler_options.GetInstructionSet();
const DexFile& dex_file = *dex_compilation_unit.GetDexFile();
uint32_t method_idx = dex_compilation_unit.GetDexMethodIndex();

//Always use the Thumb-2 assembler: some Runtime functionality
//(like implicit stack overflow checks) assume Thumb-2
DCHECK_NE(instruction_set, InstructionSet::kArm);

//Do not attempt to compile on architectures we do not support
if (!IsInstructionSetSupported(instruction_set)) {
  return nullptr;
```

```cpp
    }

    HGraph* graph = new (allocator) HGraph(
        allocator,
        arena_stack,
        dex_file,
        method_idx,
        compiler_options.GetInstructionSet(),
        kInvalidInvokeType,
        /* dead_reference_safe= */ true,
        compiler_options.GetDebuggable(),
        /* osr= */ false);

    DCHECK(Runtime::Current()->IsAotCompiler());
    DCHECK(method != nullptr);
    graph->SetArtMethod(method);

    std::unique_ptr<CodeGenerator> codegen(
        CodeGenerator::Create(graph, compiler_options,
                              compilation_stats_.get()));
    if (codegen.get() == nullptr) {
      return nullptr;
    }
    codegen->GetAssembler()->cfi().SetEnabled(
        compiler_options.GenerateAnyDebugInfo());

    PassObserver pass_observer(graph,
                               codegen.get(),
                               visualizer_output_.get(),
                               compiler_options,
                               dump_mutex_);

    {
      VLOG(compiler) << "Building intrinsic graph "
                    << pass_observer.GetMethodName();
      PassScope scope(HGraphBuilder::kBuilderPassName, &pass_observer);
      HGraphBuilder builder(graph,
          CodeItemDebugInfoAccessor(), &dex_compilation_unit,
          &dex_compilation_unit, codegen.get(), compilation_stats_.get(),
          /* interpreter_metadata= */ ArrayRef<const uint8_t>(), handles);
      builder.BuildIntrinsicGraph(method);
    }

    OptimizationDef optimizations[] = {
      //The codegen has a few assumptions that only the instruction
      //simplifier can satisfy
      OptDef(OptimizationPass::kInstructionSimplifier),
```

```
};
RunOptimizations(graph, codegen.get(), dex_compilation_unit,
                &pass_observer, handles, optimizations);

RunArchOptimizations(graph, codegen.get(), dex_compilation_unit, &pass_observer, handles);

AllocateRegisters(graph,
                codegen.get(),
                &pass_observer,
                compiler_options.GetRegisterAllocationStrategy(),
                compilation_stats_.get());
if (!codegen->IsLeafMethod()) {
  VLOG(compiler) << "Intrinsic method is not leaf: "
                << method->GetIntrinsic()
                << " " << graph->PrettyMethod();
  return nullptr;
}

codegen->Compile(code_allocator);
pass_observer.DumpDisassembly();

VLOG(compiler) << "Compiled intrinsic: "
              << method->GetIntrinsic()
              << " " << graph->PrettyMethod();
MaybeRecordStat(compilation_stats_.get(),
    MethodCompilationStat::kCompiledIntrinsic);
return codegen.release();
}
```

在OptimizingCompiler::JniCompile()函数中,在确认编译为Boot Image且要处理的方法是Intrinsic的情况下,会依次调用TryCompileIntrinsic()函数和Emit()函数。Emit()函数用于接收TryCompileIntrinsic()函数返回的CodeGenerator指针,生成一个已经编译过的方法,该编译过的方法对应了一个优化过的graph。Emit()函数的核心也是前文提到过的CompiledMethod::SwapAllocCompiledMethod()函数。Emit()函数也位于optimizing_compiler.cc中,它是OptimizingCompiler的成员函数,代码如下:

```
//第3章/optimizing_compiler.cc
CompiledMethod* OptimizingCompiler::Emit(ArenaAllocator* allocator,
    CodeVectorAllocator* code_allocator, CodeGenerator* codegen,
    const dex::CodeItem* code_item_for_osr_check) const {
  ArenaVector<linker::LinkerPatch> linker_patches =
      EmitAndSortLinkerPatches(codegen);
  ScopedArenaVector<uint8_t> stack_map =
      codegen->BuildStackMaps(code_item_for_osr_check);
```

```cpp
    CompiledMethodStorage* storage = GetCompiledMethodStorage();
    CompiledMethod* compiled_method = CompiledMethod::SwapAllocCompiledMethod(
        storage,
        codegen->GetInstructionSet(),
        code_allocator->GetMemory(),
        ArrayRef<const uint8_t>(stack_map),
        ArrayRef<const uint8_t>(*codegen->GetAssembler()->cfi().data()),
        ArrayRef<const linker::LinkerPatch>(linker_patches));

    for (const linker::LinkerPatch& patch : linker_patches) {
      if (codegen->NeedsThunkCode(patch) && storage->GetThunkCode(patch).empty()) {
        ArenaVector<uint8_t> code(allocator->Adapter());
        std::string debug_name;
        codegen->EmitThunkCode(patch, &code, &debug_name);
        storage->SetThunkCode(patch, ArrayRef<const uint8_t>(code), debug_name);
      }
    }

    return compiled_method;
}
```

到目前为止,已经完成了 OptimizingCompiler::JniCompile()函数中在确认编译为 Boot Image 且要处理的方法是 Intrinsic 的情况下的分支。接下来分析编译后不为 Boot Image 的分支。主要通过 ArtQuickJniCompileMethod() 函数和 CompiledMethod::SwapAllocCompiledMethod() 函数生成已经编译完成的方法,也就是生成一个 CompiledMethod 对象。因为 CompiledMethod::SwapAllocCompiledMethod()函数在前文已经分析过了,所以这里重点介绍 ArtQuickJniCompileMethod()函数。

ArtQuickJniCompileMethod()函数的实现位于 art/compiler/jni/quick/jni_compiler.cc 中,它主要通过指令集是 32 位还是 64 位调用对应的 ArtJniCompileMethodInternal()函数,代码如下:

```cpp
//第 3 章/jni_compiler.cc
JniCompiledMethod ArtQuickJniCompileMethod(
    const CompilerOptions& compiler_options,
    uint32_t access_flags, uint32_t method_idx,
    const DexFile& dex_file) {
  if (Is64BitInstructionSet(compiler_options.GetInstructionSet())) {
    return ArtJniCompileMethodInternal<PointerSize::k64>(
        compiler_options, access_flags, method_idx, dex_file);
  } else {
    return ArtJniCompileMethodInternal<PointerSize::k32>(
        compiler_options, access_flags, method_idx, dex_file);
  }
}
```

ArtJniCompileMethodInternal()函数也位于 art/compiler/jni/quick/jni_compiler.cc 中，它的作用是：Generate the JNI bridge for the given method, general contract:-Arguments are in the managed Runtime format, either on stack or in registers, a reference to the method object is supplied as part of this convention。ArtJniCompileMethodInternal()函数算是真正地开始对 JNI 的细节进行处理。

本部分内容对 OptimizingCompiler::JniCompile()函数中的两个分支路径分别进行了分析和跟踪，展示了 OptimizingCompiler::JniCompile()函数的主要执行流程和其中的基本环节。对于其中所涉及的构建 HGraph、运行优化（RunOptimizations()）、运行目标优化（RunArchOptimizations()）、寄存器分配和代码生成（codegen—>Compile()）等内容，会在后续章节具体展开介绍。

3.2 OptimizingCompiler::Compile()

dex2oat 对于 dex 有 3 种处理方式，前文已经对这 3 种处理方式中的 dex2dex 处理方式和与 JNI 相关的处理进行了介绍，只剩下 dex 到机器码所对应的 OptimizingCompiler::Compile()函数没有介绍。本部分将对 OptimizingCompiler::Compile()函数及其所涉及的流程进行分析。

OptimizingCompiler::Compile()函数是 OptimizingCompiler 类的一个重要的成员函数，它通过在是 Intrinsic 方法且 method 不为空的时候，执行 TryCompileIntrinsic()函数构建 CodeGenerator 指针；否则，执行 TryCompile()函数构建 CodeGenerator 指针。最后，通过 Emit()函数来生成编译完的方法。OptimizingCompiler::Compile()函数的代码如下：

```
//第3章/optimizing_compiler.cc
CompiledMethod* OptimizingCompiler::Compile(
    const dex::CodeItem* code_item, uint32_t access_flags,
    InvokeType invoke_type, uint16_t class_def_idx,
    uint32_t method_idx, Handle<mirror::ClassLoader> jclass_loader,
    const DexFile& dex_file,
    Handle<mirror::DexCache> dex_cache) const {
  const CompilerOptions& compiler_options = GetCompilerOptions();
  CompiledMethod* compiled_method = nullptr;
  Runtime* Runtime = Runtime::Current();
  DCHECK(Runtime->IsAotCompiler());
  const VerifiedMethod* verified_method =
    compiler_options.GetVerifiedMethod(&dex_file, method_idx);
  DCHECK(!verified_method->HasRuntimeThrow());
  if (compiler_options.IsMethodVerifiedWithoutFailures(method_idx,
      class_def_idx, dex_file) ||
      verifier::CanCompilerHandleVerificationFailure(
          verified_method->GetEncounteredVerificationFailures()))  {
```

```cpp
ArenaAllocator allocator(Runtime->GetArenaPool());
ArenaStack arena_stack(Runtime->GetArenaPool());
CodeVectorAllocator code_allocator(&allocator);
std::unique_ptr<CodeGenerator> codegen;
bool compiled_intrinsic = false;
{
  ScopedObjectAccess soa(Thread::Current());
  ArtMethod* method =
      Runtime->GetClassLinker()->ResolveMethod<ClassLinker::
          ResolveMode::kCheckICCEAndIAE>(
              method_idx, dex_cache, jclass_loader, /* referrer = */ nullptr, invoke_type);
  DCHECK_EQ(method == nullptr, soa.Self()->IsExceptionPending());
  soa.Self()->ClearException(); //Suppress exception if any.
  VariableSizedHandleScope handles(soa.Self());
  Handle<mirror::Class> compiling_class =
      handles.NewHandle(method != nullptr ?
      method->GetDeclaringClass() : nullptr);
  DexCompilationUnit dex_compilation_unit(
      jclass_loader,
      Runtime->GetClassLinker(),
      dex_file,
      code_item,
      class_def_idx,
      method_idx,
      access_flags,
      /* verified_method = */ nullptr,
      dex_cache,
      compiling_class);
  //Go to native so that we don't block GC during compilation
  ScopedThreadSuspension sts(soa.Self(), kNative);
  if (method != nullptr && UNLIKELY(method->IsIntrinsic())) {
    DCHECK(compiler_options.IsBootImage());
    codegen.reset(
        TryCompileIntrinsic(&allocator,
                            &arena_stack,
                            &code_allocator,
                            dex_compilation_unit,
                            method,
                            &handles));
    if (codegen != nullptr) {
      compiled_intrinsic = true;
    }
  }
  if (codegen == nullptr) {
    codegen.reset(
        TryCompile(&allocator,
```

```cpp
                        &arena_stack,
                        &code_allocator,
                        dex_compilation_unit,
                        method,
                        compiler_options.IsBaseline(),
                        /* osr= */ false,
                        &handles));
    }
  }
  if (codegen.get() != nullptr) {
    compiled_method = Emit(&allocator,
                           &code_allocator,
                           codegen.get(),
                           compiled_intrinsic ? nullptr : code_item);
    if (compiled_intrinsic) {
      compiled_method->MarkAsIntrinsic();
    }

    if (kArenaAllocatorCountAllocations) {
      codegen.reset();
      size_t total_allocated = allocator.BytesAllocated() +
          arena_stack.PeakBytesAllocated();
      if (total_allocated > kArenaAllocatorMemoryReportThreshold) {
        MemStats mem_stats(allocator.GetMemStats());
        MemStats peak_stats(arena_stack.GetPeakStats());
        LOG(INFO) << "Used " << total_allocated
                  << " Bytes of arena memory for compiling "
                  << dex_file.PrettyMethod(method_idx)
                  << "\n" << Dumpable<MemStats>(mem_stats)
                  << "\n" << Dumpable<MemStats>(peak_stats);
      }
    }
  }
} else {
  MethodCompilationStat method_stat;
  if (compiler_options.VerifyAtRuntime()) {
    method_stat = MethodCompilationStat::kNotCompiledVerifyAtRuntime;
  } else {
    method_stat = MethodCompilationStat::kNotCompiledVerificationError;
  }
  MaybeRecordStat(compilation_stats_.get(), method_stat);
}

if (kIsDebugBuild &&
    compiler_options.CompilingWithCoreImage() &&
    IsInstructionSetSupported(compiler_options.GetInstructionSet())) {
```

```
        //For testing purposes, we put a special marker on method names
        //that should be compiled with this compiler (when the
        //instruction set is supported). This makes sure we're not
        //regressing
        std::string method_name = dex_file.PrettyMethod(method_idx);
        bool shouldCompile = method_name.find(" $ opt $ ") != std::string::npos;
        DCHECK((compiled_method != nullptr) || !shouldCompile)
            << "Didn't compile " << method_name;
    }

    return compiled_method;
}
```

在这里,代码中在判断条件的时候使用了 UNLIKELY 宏,读者需要明确 UNLIKELY 宏并不会改变判断条件,只是预判该条件是否为小概率成功。从代码 UNLIKELY(method->IsIntrinsic())可以得知,在 dex 转化为 oat 的过程中,要处理的 Intrinsic 方法是少数,要处理的普通方法才是大多数情况,而这时候,处理 Intrinsic 方法调用的是 TryCompileIntrinsic()函数,而处理普通方法调用的是 TryCompile()函数。其中,TryCompileIntrinsic()函数在前文 OptimizingCompiler::JniCompile()函数的分析时已经介绍过了,后续只分析 TryCompile()函数。

TryCompile()函数和 TryCompileIntrinsic()函数的基本流程是一致的,都是通过构建 HGraph、运行优化(RunBaselineOptimizations()或 RunOptimizations())、分配寄存器(AllocateRegisters())、代码生成(codegen—>Compile())等几个环节的处理,最终返回一个 CodeGenerator 指针,二者不同的地方在于执行的优化内容不同,即在运行优化环节的内容不同。TryCompile()函数的代码如下:

```
//第 3 章/optimizing_compiler.cc
CodeGenerator* OptimizingCompiler::TryCompile(
    ArenaAllocator* allocator, ArenaStack* arena_stack,
    CodeVectorAllocator* code_allocator,
    const DexCompilationUnit& dex_compilation_unit,
    ArtMethod* method, bool baseline, bool osr,
    VariableSizedHandleScope* handles) const {
  MaybeRecordStat(compilation_stats_.get(), MethodCompilationStat::kAttemptBytecodeCompilation);
  const CompilerOptions& compiler_options = GetCompilerOptions();
  InstructionSet instruction_set = compiler_options.GetInstructionSet();
  const DexFile& dex_file =  * dex_compilation_unit.GetDexFile();
  uint32_t method_idx = dex_compilation_unit.GetDexMethodIndex();
  const dex::CodeItem* code_item = dex_compilation_unit.GetCodeItem();

  //Always use the Thumb-2 assembler: some Runtime functionality
  //(like implicit stack overflow checks) assume Thumb-2
```

```cpp
DCHECK_NE(instruction_set, InstructionSet::kArm);

//Do not attempt to compile on architectures we do not support
if (!IsInstructionSetSupported(instruction_set)) {
  MaybeRecordStat(compilation_stats_.get(),
      MethodCompilationStat::kNotCompiledUnsupportedIsa);
  return nullptr;
}

if (Compiler::IsPathologicalCase(*code_item, method_idx, dex_file)) {
  MaybeRecordStat(compilation_stats_.get(), MethodCompilationStat::kNotCompiledPathological);
  return nullptr;
}

//Implementation of the space filter: do not compile a code item whose size in code units is
//bigger than 128
static constexpr size_t kSpaceFilterOptimizingThreshold = 128;
if ((compiler_options.GetCompilerFilter() == CompilerFilter::kSpace)
    && (CodeItemInstructionAccessor(dex_file, code_item).InsnsSizeInCodeUnits() >
    kSpaceFilterOptimizingThreshold)) {
  MaybeRecordStat(compilation_stats_.get(), MethodCompilationStat::kNotCompiledSpaceFilter);
  return nullptr;
}

CodeItemDebugInfoAccessor code_item_accessor(dex_file, code_item, method_idx);

bool dead_reference_safe;
ArrayRef<const uint8_t> interpreter_metadata;
//For AOT compilation, we may not get a method, for example if its
//class is erroneous, possibly due to an unavailable superclass. JIT
//should always have a method
DCHECK(Runtime::Current()->IsAotCompiler() || method != nullptr);
if (method != nullptr) {
  const dex::ClassDef* containing_class;
  {
    ScopedObjectAccess soa(Thread::Current());
    containing_class = &method->GetClassDef();
    interpreter_metadata = method->GetQuickenedInfo();
  }
  //MethodContainsRSensitiveAccess is currently slow, but
  //HasDeadReferenceSafeAnnotation() is currently rarely true
  dead_reference_safe =
      annotations::HasDeadReferenceSafeAnnotation(dex_file, *containing_class)
      && !annotations::MethodContainsRSensitiveAccess(dex_file, *containing_class, method_idx);
} else {
  //If we could not resolve the class, conservatively assume it's
```

```
    //dead - reference unsafe
    dead_reference_safe = false;
}

HGraph* graph = new (allocator) HGraph(
    allocator,
    arena_stack,
    dex_file,
    method_idx,
    compiler_options.GetInstructionSet(),
    kInvalidInvokeType,
    dead_reference_safe,
    compiler_options.GetDebuggable(),
    /* osr= */ osr);

if (method != nullptr) {
  graph->SetArtMethod(method);
}

std::unique_ptr<CodeGenerator> codegen(
    CodeGenerator::Create(graph,
                          compiler_options,
                          compilation_stats_.get()));
if (codegen.get() == nullptr) {
  MaybeRecordStat(compilation_stats_.get(), MethodCompilationStat::kNotCompiledNoCodegen);
  return nullptr;
}
codegen->GetAssembler()->cfi().SetEnabled(compiler_options.
    GenerateAnyDebugInfo());

PassObserver pass_observer(graph,
                           codegen.get(),
                           visualizer_output_.get(),
                           compiler_options,
                           dump_mutex_);

{
  VLOG(compiler) << "Building " << pass_observer.GetMethodName();
  PassScope scope(HGraphBuilder::kBuilderPassName, &pass_observer);
  HGraphBuilder builder(graph,
                        code_item_accessor,
                        &dex_compilation_unit,
                        &dex_compilation_unit,
                        codegen.get(),
                        compilation_stats_.get(),
                        interpreter_metadata,
```

```cpp
                              handles);
  GraphAnalysisResult result = builder.BuildGraph();
  if (result != kAnalysisSuccess) {
    switch (result) {
      case kAnalysisSkipped: {
        MaybeRecordStat(compilation_stats_.get(),
            MethodCompilationStat::kNotCompiledSkipped);
        break;
      }
      case kAnalysisInvalidBytecode: {
        MaybeRecordStat(compilation_stats_.get(),
            MethodCompilationStat::kNotCompiledInvalidBytecode);
        break;
      }
      case kAnalysisFailThrowCatchLoop: {
        MaybeRecordStat(compilation_stats_.get(),
            MethodCompilationStat::kNotCompiledThrowCatchLoop);
        break;
      }
      case kAnalysisFailAmbiguousArrayOp: {
        MaybeRecordStat(compilation_stats_.get(),
           MethodCompilationStat::kNotCompiledAmbiguousArrayOp);
        break;
      }
      case kAnalysisFailIrreducibleLoopAndStringInit: {
        MaybeRecordStat(compilation_stats_.get(),
            MethodCompilationStat::
            kNotCompiledIrreducibleLoopAndStringInit);
        break;
      }
      case kAnalysisSuccess:
        UNREACHABLE();
    }
    pass_observer.SetGraphInBadState();
    return nullptr;
  }
}

if (baseline) {
  RunBaselineOptimizations(graph, codegen.get(), dex_compilation_unit, &pass_observer,
handles);
} else {
  RunOptimizations(graph, codegen.get(), dex_compilation_unit, &pass_observer, handles);
}

RegisterAllocator::Strategy regalloc_strategy =
```

```
                compiler_options.GetRegisterAllocationStrategy();
        AllocateRegisters(graph,
                        codegen.get(),
                        &pass_observer,
                        regalloc_strategy,
                        compilation_stats_.get());

        codegen->Compile(code_allocator);
        pass_observer.DumpDisassembly();

        MaybeRecordStat(compilation_stats_.get(), MethodCompilationStat::kCompiledBytecode);
        return codegen.release();
    }
```

TryCompile()函数里面的核心内容还是构建 HGraph、优化、寄存器分配、代码生成等环节,这些环节都是编译过程中的重要环节,在后续的内容中会对这些环节进行详细介绍。

本部分内容介绍了 dex2oat 是 dex 处理中最普通的一种,它可将 dex 转换为 oat 格式,和之前介绍的 dex2dex 处理及 JNI 相关处理一起构成了 dex2oat 的 3 种处理方式。这个处理是通过 OptimizingCompiler::Compile()函数实现的,而 OptimizingCompiler::Compile()函数的实现则是通过 TryCompile()函数和 TryCompileIntrinsic()函数拆分成了几个常用的编译环节进行处理的,这几个环节将在后续章节逐个进行介绍。

3.3　HGraph 的构建

前文提到 HGraph 构建等几个环节是 OptimizingCompiler 中的重要组成部分,也是 dex2oat 工具中的重要环节,本部分将对 HGraph 构建展开介绍。

HGraph 构建的主要内容是构建基本代码块并为其构建支配树,即构建 SSA 形式的树。它的具体实现位于 HGraphBuilder 类中。HGraphBuilder 类的定义和实现位于 art/compiler/optimizing/目录下的 builder.h 和 builder.cc 中。

HGraph 构建的入口在 HGraphBuilder::BuildGraph()函数中,它在 OptimizingCompiler::TryCompile()函数和 OptimizingCompiler::TryCompileIntrinsic()函数中被调用。HGraphBuilder::BuildGraph()函数的实现位于 builder.cc 中,代码如下:

```
//第 3 章/builder.cc
GraphAnalysisResult HGraphBuilder::BuildGraph() {
    DCHECK(code_item_accessor_.HasCodeItem());
    DCHECK(graph_->GetBlocks().empty());

    graph_->SetNumberOfVRegs(code_item_accessor_.RegistersSize());
    graph_->SetNumberOfInVRegs(code_item_accessor_.InsSize());
```

```cpp
graph_->SetMaximumNumberOfOutVRegs(code_item_accessor_.OutsSize());
graph_->SetHasTryCatch(code_item_accessor_.TriesSize() != 0);

//Use ScopedArenaAllocator for all local allocations
ScopedArenaAllocator local_allocator(graph_->GetArenaStack());
HBasicBlockBuilder block_builder(graph_, dex_file_,
                                 code_item_accessor_,
                                 &local_allocator);
SsaBuilder ssa_builder(graph_,
                       dex_compilation_unit_->GetClassLoader(),
                       dex_compilation_unit_->GetDexCache(),
                       handles_, &local_allocator);
HInstructionBuilder instruction_builder(graph_,
                                        &block_builder,
                                        &ssa_builder,
                                        dex_file_,
                                        code_item_accessor_,
                                        return_type_,
                                        dex_compilation_unit_,
                                        outer_compilation_unit_,
                                        code_generator_,
                                        interpreter_metadata_,
                                        compilation_stats_,
                                        handles_,
                                        &local_allocator);

//(1) Create basic blocks and link them together. Basic blocks are left
// unpopulated with the exception of synthetic blocks, e.g.
// HTryBoundaries
if (!block_builder.Build()) {
  return kAnalysisInvalidBytecode;
}

//(2) Decide whether to skip this method based on its code size and
// number of branches
if (SkipCompilation(block_builder.GetNumberOfBranches())) {
  return kAnalysisSkipped;
}

//(3) Build the dominator tree and fill in loop and try/catch metadata
GraphAnalysisResult result = graph_->BuildDominatorTree();
if (result != kAnalysisSuccess) {
  return result;
}

//(4) Populate basic blocks with instructions
```

```
  if (!instruction_builder.Build()) {
    return kAnalysisInvalidBytecode;
  }

  //(5) Type the graph and eliminate dead/redundant phis
  return ssa_builder.BuildSsa();
}
```

BuildGraph()函数中有几个关键的操作需要关注,依次为构建基于基本代码块的CFG,它对应的代码为 block_builder.Build()函数;构建支配树,它对应的代码为 graph_->BuildDominatorTree()函数;构建 SSA,它对应的代码为 ssa_builder.BuildSsa()函数。在本节的后续部分,将对这几个操作进行介绍。

3.3.1 构建基于基本代码块的 CFG

BuildGraph()函数中调用了 block_builder.Build()函数,它实现了构建基于基本代码块的 CFG 操作,它是 HBasicBlockBuilder 类的 Build()函数。

在调用 Build()函数之前,需要构建一个 HBasicBlockBuilder 类的对象,这点在 HGraphBuilder::BuildGraph()函数中有体现,具体实现的代码如下:

```
HBasicBlockBuilder block_builder(graph_, dex_file_,
                                 code_item_accessor_,
                                 &local_allocator);
```

在此之后,才调用了 block_builder.Build()函数。

HBasicBlockBuilder 类的定义和实现位于 art/compiler/optimizing/目录下的 block_builder.h 和 block_builder.cc 中。跳过 HBasicBlockBuilder 类的构造函数,直接看关键的 Build()函数。HBasicBlockBuilder 类的 Build()函数实现位于 block_builder.cc 中,代码如下:

```
//第3章/block_builder.cc
bool HBasicBlockBuilder::Build() {
  DCHECK(code_item_accessor_.HasCodeItem());
  DCHECK(graph_->GetBlocks().empty());

  graph_->SetEntryBlock(new (allocator_) HBasicBlock(graph_, kNoDexPc));
  graph_->SetExitBlock(new (allocator_) HBasicBlock(graph_, kNoDexPc));

  //TODO(dbrazdil): Do CreateBranchTargets and ConnectBasicBlocks in one
  //pass
  if (!CreateBranchTargets()) {
    return false;
  }
```

```
    ConnectBasicBlocks();
    InsertTryBoundaryBlocks();

    return true;
}
```

Build 方法的核心内容是调用了 CreateBranchTargets（）函数、ConnectBasicBlocks（）函数和 InsertTryBoundaryBlocks（）函数。其中，CreateBranchTargets（）函数和 ConnectBasicBlocks（）函数用于创建基本代码块并把这些基本代码块链接起来，这时候已基本构建了基于基本代码块的 CFG；InsertTryBoundaryBlocks（）函数处理了与 try 和 catch 相关的基本代码块。

3.3.2　构建支配树

支配树（Dominator Tree）是一种树状控制流图，它的所有节点的子节点是被该节点最近支配的节点，但是，反过来，节点对于其子节点在控制流中并不一定是最近必经节点。一个节点的最近必经节点是唯一的，所以这些节点可以组成一个树状图，开始节点即为树根。以图 3.1 的控制流图为例，下面对其进行简单介绍。

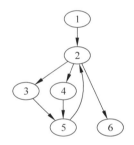

图 3.1　控制流图

（图源：·https://en.wikipedia.org/wiki/Dominator_（graph_theory））

在图 3.1 中，1 可以支配 2，2 可以支配 3、4、6，3、4 都可以支配 5，但只有 2 是 5 的最近必经节点，3 和 4 不是。

BuildGraph（）函数中调用了 graph_ -> BuildDominatorTree（）函数，它进行的是构建支配树的动作，这实际上调用的是 HGraph::BuildDominatorTree（）函数，HGraph::BuildDominatorTree（）函数的实现位于 art/compiler/optimizing/nodes.cc 中，代码如下：

```
//第 3 章/nodes.cc
GraphAnalysisResult HGraph::BuildDominatorTree() {
    //Allocate memory from local ScopedArenaAllocator
    ScopedArenaAllocator allocator(GetArenaStack());

    ArenaBitVector visited(&allocator, blocks_.size(), false, kArenaAllocGraphBuilder);
    visited.ClearAllBits();
```

```
    //(1) Find the back edges in the graph doing a DFS traversal
    FindBackEdges(&visited);

    //(2) Remove instructions and phis from blocks not visited during
    //   the initial DFS as users from other instructions, so that
    //   users can be safely removed before uses later
    RemoveInstructionsAsUsersFromDeadBlocks(visited);

    //(3) Remove blocks not visited during the initial DFS
    //   Step (5) requires dead blocks to be removed from the
    //   predecessors list of live blocks
    RemoveDeadBlocks(visited);

    //(4) Simplify the CFG now, so that we don't need to recompute
    //   dominators and the reverse post order
    SimplifyCFG();

    //(5) Compute the dominance information and the reverse post order
    ComputeDominanceInformation();

    //(6) Analyze loops discovered through back edge analysis, and
    //   set the loop information on each block
    GraphAnalysisResult result = AnalyzeLoops();
    if (result != kAnalysisSuccess) {
      return result;
    }

    //(7) Precompute per-block try membership before entering the SSA
    //builder, which needs the information to build catch block phis from
    //values of locals at throwing instructions inside try blocks
    ComputeTryBlockInformation();

    return kAnalysisSuccess;
}
```

HGraph::BuildDominatorTree()函数的实现过程被分为了7个部分,其中核心部分是第(5)步,它调用了ComputeDominanceInformation()函数,进行支配信息和逆后序的计算。逆后序是接近前序但和前序不同的一种访问方式,其访问的时候,在访问一个节点之前,需要访问该节点的所有前序节点,在数据流分析中比较常见。HGraph::BuildDominatorTree()函数的具体代码如下：

```
//第3章/nodes.cc
void HGraph::ComputeDominanceInformation() {
  DCHECK(reverse_post_order_.empty());
```

```cpp
reverse_post_order_.reserve(blocks_.size());
reverse_post_order_.push_back(entry_block_);

//Allocate memory from local ScopedArenaAllocator
ScopedArenaAllocator allocator(GetArenaStack());
//Number of visits of a given node, indexed by block id
ScopedArenaVector<size_t> visits(blocks_.size(), 0u,
    allocator.Adapter(kArenaAllocGraphBuilder));
//Number of successors visited from a given node, indexed by block id
ScopedArenaVector<size_t> successors_visited(blocks_.size(),
    0u, allocator.Adapter(kArenaAllocGraphBuilder));
//Nodes for which we need to visit successors
ScopedArenaVector<HBasicBlock*>
    worklist(allocator.Adapter(kArenaAllocGraphBuilder));
constexpr size_t kDefaultWorklistSize = 8;
worklist.reserve(kDefaultWorklistSize);
worklist.push_back(entry_block_);

while (!worklist.empty()) {
  HBasicBlock* current = worklist.back();
  uint32_t current_id = current->GetBlockId();
  if (successors_visited[current_id] ==
      current->GetSuccessors().size()) {
    worklist.pop_back();
  } else {
    HBasicBlock* successor =
        current->GetSuccessors()[successors_visited[current_id]++];
    UpdateDominatorOfSuccessor(current, successor);

    //Once all the forward edges have been visited, we know the
    //immediate dominator of the block. We can then start visiting its
    //successors
    if (++visits[successor->GetBlockId()] ==
        successor->GetPredecessors().size() - successor->NumberOfBackEdges()) {
      reverse_post_order_.push_back(successor);
      worklist.push_back(successor);
    }
  }
}

//Check if the graph has back edges not dominated by their respective
//headers. If so, we need to update the dominators of those headers
//and recursively of their successors. We do that with a fix-point
//iteration over all blocks. The algorithm is guaranteed to terminate
//because it loops only if the sum of all dominator chains has
//decreased in the current iteration
```

```cpp
    bool must_run_fix_point = false;
    for (HBasicBlock* block : blocks_) {
      if (block != nullptr &&
          block->IsLoopHeader() &&
          block->GetLoopInformation()->HasBackEdgeNotDominatedByHeader()) {
        must_run_fix_point = true;
        break;
      }
    }
    if (must_run_fix_point) {
      bool update_occurred = true;
      while (update_occurred) {
        update_occurred = false;
        for (HBasicBlock* block : GetReversePostOrder()) {
          for (HBasicBlock* successor : block->GetSuccessors()) {
            update_occurred |= UpdateDominatorOfSuccessor(block, successor);
          }
        }
      }
    }

    //Make sure that there are no remaining blocks whose dominator
    //information needs to be updated
    if (kIsDebugBuild) {
      for (HBasicBlock* block : GetReversePostOrder()) {
        for (HBasicBlock* successor : block->GetSuccessors()) {
          DCHECK(!UpdateDominatorOfSuccessor(block, successor));
        }
      }
    }

    //Populate `dominated_blocks_` information after computing all
    //dominators. The potential presence of irreducible loops requires to
    //do it after
    for (HBasicBlock* block : GetReversePostOrder()) {
      if (!block->IsEntryBlock()) {
        block->GetDominator()->AddDominatedBlock(block);
      }
    }
}
```

这里计算出来的支配信息用于后续构建 SSA 形式的树，包括对于 phi 的优化都会用到。图的逆后序（Reverse Post Order，RPO）遍历在执行迭代算法时非常有效，它在访问节点之前会尽可能多地访问该节点的前驱节点，这一点需要和前序遍历区分开来。

3.3.3 构建 SSA

BuildGraph()函数中调用了 ssa_builder.BuildSsa()函数,它是 BuildGraph()函数中的最后一个操作,它所进行的是构建 SSA 的操作。SSA 是静态单赋值形式(Static Single Assignment Form)的简称,它是一种中间表现形式(Intermediate Representation,IR),主要特点是每个变量只被赋值一次,这种中间表现形式在编译器中使用得非常多。

ssa_builder 是一个 SsaBuilder 类的对象,它的定义也位于 BuildGraph()函数中,所以 ssa_builder.BuildSsa()函数具体执行的是 SsaBuilder::BuildSsa()函数。SsaBuilder 类的定义和实现位于 art/compiler/optimizing 目录下的 ssa_builder.h 和 ssa_builder.cc 中。SsaBuilder::BuildSsa()函数的实现位于 ssa_builder.cc 中,代码如下:

```
//第3章/ssa_builder.cc
GraphAnalysisResult SsaBuilder::BuildSsa() {
  DCHECK(!graph_->IsInSsaForm());

  //Propagate types of phis. At this point, phis are typed void in the
  //general case, or float/double/reference if we created an equivalent
  //phi. So we need to propagate the types across phis to give them a
  //correct type. If a type conflict is detected in this stage, the phi
  //is marked dead
  RunPrimitiveTypePropagation();

  //Now that the correct primitive types have been assigned, we can get
  //rid of redundant phis. Note that we cannot do this phase before type
  //propagation, otherwise we could get rid of phi equivalents, whose
  //presence is a requirement for the type propagation phase. Note that
  //this is to satisfy statement (a) of the SsaBuilder (see
  //ssa_builder.h)
  SsaRedundantPhiElimination(graph_).Run();

  //Fix the type for null constants which are part of an equality
  //comparison. We need to do this after redundant phi elimination, to
  //ensure the only cases that we can see are reference comparison
  //against 0. The redundant phi elimination ensures we do not see a phi
  //taking two 0 constants in a HEqual or HNotEqual
  FixNullConstantType();

  //Compute type of reference type instructions. The pass assumes that
  //NullConstant has been fixed up
  ReferenceTypePropagation(graph_,
                           class_loader_,
                           dex_cache_,
                           handles_,
```

```cpp
                            /* is_first_run= */ true).Run();

//HInstructionBuilder duplicated ArrayGet instructions with ambiguous
//type (int/float or long/double) and marked ArraySets with ambiguous
//input type. Now that RTP computed the type of the array input, the
//ambiguity can be resolved and the correct equivalents kept
if (!FixAmbiguousArrayOps()) {
  return kAnalysisFailAmbiguousArrayOp;
}

//Mark dead phis. This will mark phis which are not used by
//instructions or other live phis. If compiling as debuggable code
//phis will also be kept live if they have an environment use
SsaDeadPhiElimination dead_phi_elimination(graph_);
dead_phi_elimination.MarkDeadPhis();

//Make sure environments use the right phi equivalent: a phi marked
//dead can have a phi equivalent that is not dead. In that case we
//have to replace it with the live equivalent because deoptimization
//and try/catch rely on environments containing values of all live
//vregs at that point. Note that there can be multiple phis for the
//same Dex register that are live (for example when merging
//constants), in which case it is okay for the environments to just
//reference one
FixEnvironmentPhis();

//Now that the right phis are used for the environments, we can
//eliminate phis we do not need. Regardless of the debuggable status
//this phase is necessary for statement (b) of the SsaBuilder (see
//ssa_builder.h), as well as for the code generation, which does not
//deal with phis of conflicting input types
dead_phi_elimination.EliminateDeadPhis();

//Replace Phis that feed in a String.<init> during instruction
//building. We run this after redundant and dead phi elimination to
//make sure the phi will have been replaced by the actual allocation
//Only with an irreducible loop a phi can still be the input, in which
//case we bail
if (!ReplaceUninitializedStringPhis()) {
  return kAnalysisFailIrreducibleLoopAndStringInit;
}

//HInstructionBuidler replaced uses of NewInstances of String with the
//results of their corresponding StringFactory calls. Unless the
//String objects are used before they are initialized, they can be
//replaced with NullConstant. Note that this optimization is valid
```

```
//only if unsimplified code does not use the uninitialized value
//because we assume execution can be deoptimized at any safepoint. We
//must therefore perform it before any other optimizations
RemoveRedundantUninitializedStrings();

graph_->SetInSsaForm();
return kAnalysisSuccess;
}
```

构建 SSA 是在之前构建的 CFG 基础之上，清理掉重复和不使用的 phi 指令，最终完成一个简洁的 SSA，为后续的优化做准备。

本节介绍了 HGraph 的主要内容，并且对其中的构建基于基本代码块的 CFG、构建支配树和构建 SSA 等内容进行了介绍。

3.4 优化

OptimizingCompiler 包含了基于 ART IR HInstruction 的优化，这位于 HGraph 构建环节之后，也是以 pass 的形式逐个多次运行，以便进行优化。本部分内容将对优化的整体情况进行介绍。

pass 在中文有不同的翻译版本，有翻译为"趟"的，也有翻译为"通道"的，其运作机制类似于一个滤网，对于输入内容中的特定内容进行处理，然后输出结果。为了避免产生歧义，本书中后续部分仍然使用 pass 这个称呼。

前文对于 TryCompile() 函数和 TryCompileIntrinsic() 函数进行过分析，涉及了关于优化的几条路径。对 ART IR 进行的优化，主要路径还是 RunOptimizations() 函数。在这里，将 RunOptimizations() 函数作为 ART IR 进行优化的入口函数进行分析，它位于 art/compiler/optimizing/optimizing_compiler.cc 文件中，该函数为重载函数，有 5 个参数、6 个参数和 7 个参数的不同版本，代码如下：

```
//第 3 章/optimizing_compiler.cc
void RunOptimizations(HGraph* graph,
    CodeGenerator* codegen,
    const DexCompilationUnit& dex_compilation_unit,
    PassObserver* pass_observer,
    VariableSizedHandleScope* handles) const;

template<size_t length> bool RunOptimizations(
    HGraph* graph,
    CodeGenerator* codegen,
    const DexCompilationUnit& dex_compilation_unit,
    PassObserver* pass_observer,
```

```
        VariableSizedHandleScope* handles,
        const OptimizationDef (&definitions)[length]) const;

bool RunOptimizations(HGraph* graph,
    CodeGenerator* codegen,
    const DexCompilationUnit& dex_compilation_unit,
    PassObserver* pass_observer,
    VariableSizedHandleScope* handles,
    const OptimizationDef definitions[],
    size_t length) ;
```

其中 5 个参数版本的 RunOptimizations() 函数在实际的执行过程中,会通过 6 个参数版本的 RunOptimizations() 函数,最终调用 7 个参数版本的 RunOptimizations() 函数,也就是说最终的优化过程实际执行是 7 个参数版本的 RunOptimizations() 函数,所以分析优化部分的实现,要从 5 个参数版本的 RunOptimizations() 函数入手。5 个参数版本的 RunOptimizations() 函数的代码如下:

```
//第 3 章/optimizing_compiler.cc
void OptimizingCompiler::RunOptimizations(HGraph* graph,
    CodeGenerator* codegen,
    const DexCompilationUnit& dex_compilation_unit,
    PassObserver* pass_observer,
    VariableSizedHandleScope* handles) const {
  const std::vector<std::string>* pass_names =
    GetCompilerOptions().GetPassesToRun();
  if (pass_names != nullptr) {
    //If passes were defined on command-line, build the optimization
    //passes and run these instead of the built-in optimizations
    //TODO: a way to define depends_on via command-line
    const size_t length = pass_names->size();
    std::vector<OptimizationDef> optimizations;
    for (const std::string& pass_name : *pass_names) {
      std::string opt_name = ConvertPassNameToOptimizationName(pass_name);
      optimizations.push_back(OptDef(OptimizationPassByName(opt_name),
                                    pass_name.c_str()));
    }
    RunOptimizations(graph,
                codegen,
                dex_compilation_unit,
                pass_observer,
                handles,
                optimizations.data(),
                length);
    return;
```

```cpp
}

OptimizationDef optimizations[] = {
    //Initial optimizations
    OptDef(OptimizationPass::kConstantFolding),
    OptDef(OptimizationPass::kInstructionSimplifier),
    OptDef(OptimizationPass::kDeadCodeElimination,
           "dead_code_elimination$initial"),
    //Inlining
    OptDef(OptimizationPass::kInliner),
    //Simplification (only if inlining occurred)
    OptDef(OptimizationPass::kConstantFolding,
           "constant_folding$after_inlining",
           OptimizationPass::kInliner),
    OptDef(OptimizationPass::kInstructionSimplifier,
           "instruction_simplifier$after_inlining",
           OptimizationPass::kInliner),
    OptDef(OptimizationPass::kDeadCodeElimination,
           "dead_code_elimination$after_inlining",
           OptimizationPass::kInliner),
    //GVN
    OptDef(OptimizationPass::kSideEffectsAnalysis,
           "side_effects$before_gvn"),
    OptDef(OptimizationPass::kGlobalValueNumbering),
    //Simplification (TODO: only if GVN occurred)
    OptDef(OptimizationPass::kSelectGenerator),
    OptDef(OptimizationPass::kConstantFolding,
           "constant_folding$after_gvn"),
    OptDef(OptimizationPass::kInstructionSimplifier,
           "instruction_simplifier$after_gvn"),
    OptDef(OptimizationPass::kDeadCodeElimination,
           "dead_code_elimination$after_gvn"),
    //High-level optimizations
    OptDef(OptimizationPass::kSideEffectsAnalysis,
           "side_effects$before_licm"),
    OptDef(OptimizationPass::kInvariantCodeMotion),
    OptDef(OptimizationPass::kInductionVarAnalysis),
    OptDef(OptimizationPass::kBoundsCheckElimination),
    OptDef(OptimizationPass::kLoopOptimization),
    //Simplification
    OptDef(OptimizationPass::kConstantFolding,
           "constant_folding$after_bce"),
    OptDef(OptimizationPass::kInstructionSimplifier,
           "instruction_simplifier$after_bce"),
    //Other high-level optimizations
    OptDef(OptimizationPass::kSideEffectsAnalysis,
```

```
            "side_effects $ before_lse"),
    OptDef(OptimizationPass::kLoadStoreAnalysis),
    OptDef(OptimizationPass::kLoadStoreElimination),
    OptDef(OptimizationPass::kCHAGuardOptimization),
    OptDef(OptimizationPass::kDeadCodeElimination,
            "dead_code_elimination $ final"),
    OptDef(OptimizationPass::kCodeSinking),
    //The codegen has a few assumptions that only the instruction
    //simplifier can satisfy. For example, the code generator does not
    //expect to see a HTypeConversion from a type to the same type
    OptDef(OptimizationPass::kInstructionSimplifier,
            "instruction_simplifier $ before_codegen"),
    //Eliminate constructor fences after code sinking to avoid
    //complicated sinking logic to split a fence with many inputs
    OptDef(OptimizationPass::kConstructorFenceRedundancyElimination)
  };
  RunOptimizations(graph,
                   codegen,
                   dex_compilation_unit,
                   pass_observer,
                   handles,
                   optimizations);

  RunArchOptimizations(graph, codegen, dex_compilation_unit,
                       pass_observer, handles);
}
```

在这里，5个参数的RunOptimizations()函数，会被转化成为6个参数的RunOptimizations()函数，或者转化为6个参数的RunOptimizations()函数和RunArchOptimizations()函数的组合。前者主要发生在5个参数的RunOptimizations()函数可以从GetCompilerOptions().GetPassesToRun()函数中获取要执行的pass列表；后者则需要5个参数的RunOptimizations()函数自己去构建并优化pass数组OptimizationDef optimizations[]，构建完成后去将optimizations传递给6个参数的RunOptimizations()函数，并且在其后执行RunArchOptimizations()函数，以便运行与架构有关的优化。其中，数组OptimizationDef optimizations[]中的具体优化会在后续的内容去进行具体介绍，这里不做具体分析。

6个参数的RunOptimizations()函数被调用后，会将参数进行展开，然后调用7个参数的RunOptimizations()函数，代码如下：

```
//第3章/optimizing_compiler.cc
template < size_t length > bool RunOptimizations(
    HGraph * graph,
    CodeGenerator * codegen,
    const DexCompilationUnit& dex_compilation_unit,
```

```
        PassObserver* pass_observer,
        VariableSizedHandleScope* handles,
        const OptimizationDef (&definitions)[length]) const {
    return RunOptimizations(graph, codegen, dex_compilation_unit,
                            pass_observer, handles, definitions, length);
}
```

7个参数的RunOptimizations()函数最终将优化的pass都展开进行运行,代码如下:

```
//第3章/optimizing_compiler.cc
bool RunOptimizations(HGraph* graph,
                     CodeGenerator* codegen,
                     const DexCompilationUnit& dex_compilation_unit,
                     PassObserver* pass_observer,
                     VariableSizedHandleScope* handles,
                     const OptimizationDef definitions[],
                     size_t length) const {
    //Convert definitions to optimization passes
    ArenaVector<HOptimization*> optimizations = ConstructOptimizations(
        definitions,
        length,
        graph->GetAllocator(),
        graph,
        compilation_stats_.get(),
        codegen,
        dex_compilation_unit,
        handles);
    DCHECK_EQ(length, optimizations.size());
    //Run the optimization passes one by one. Any "depends_on" pass
    //refers back to the most recent occurrence of that pass, skipped or
    //executed
    std::bitset<static_cast<size_t>(OptimizationPass::kLast) + 1u> pass_changes;
    pass_changes[static_cast<size_t>(OptimizationPass::kNone)] = true;
    bool change = false;
    for (size_t i = 0; i < length; ++i) {
        if (pass_changes[static_cast<size_t>(definitions[i].depends_on)]) {
            //Execute the pass and record whether it changed anything
            PassScope scope(optimizations[i]->GetPassName(), pass_observer);
            bool pass_change = optimizations[i]->Run();
            pass_changes[static_cast<size_t>(definitions[i].pass)] = pass_change;
            if (pass_change) {
                change = true;
            } else {
                scope.SetPassNotChanged();
            }
        }
```

```
        } else {
          //Skip the pass and record that nothing changed
          pass_changes[static_cast<size_t>(definitions[i].pass)] = false;
        }
      }
      return change;
    }
```

优化 pass 的运行具体是通过调用该 pass 的 Run() 函数实现的,在代码中是通过 optimizations[i]->Run()实现的。

此外,5 个参数的 RunOptimizations() 函数,还调用了 RunArchOptimizations() 函数。RunArchOptimizations() 函数负责运行一些与具体目标架构相关的优化,它针对 Arm、Arm64、Mips、Mips64、X86 和 X86_64 为每个目标平台准备了特有的一些优化,通过调用 6 个参数的 RunOptimizations() 函数,最终通过 7 个参数的 RunOptimizations() 函数去执行优化后的 pass,代码如下:

```
//第 3 章/optimizing_compiler.cc
bool OptimizingCompiler::RunArchOptimizations(HGraph* graph,
    CodeGenerator* codegen,
    const DexCompilationUnit& dex_compilation_unit,
    PassObserver* pass_observer,
    VariableSizedHandleScope* handles) const {
  switch (codegen->GetCompilerOptions().GetInstructionSet()) {
#if defined(ART_ENABLE_CODEGEN_arm)
    case InstructionSet::kThumb2:
    case InstructionSet::kArm: {
      OptimizationDef arm_optimizations[] = {
        OptDef(OptimizationPass::kInstructionSimplifierArm),
        OptDef(OptimizationPass::kSideEffectsAnalysis),
        OptDef(OptimizationPass::kGlobalValueNumbering, "GVN$after_arch"),
        OptDef(OptimizationPass::kScheduling)
      };
      return RunOptimizations(graph,
                              codegen,
                              dex_compilation_unit,
                              pass_observer,
                              handles,
                              arm_optimizations);
    }
#endif
#ifdef ART_ENABLE_CODEGEN_arm64
    case InstructionSet::kArm64: {
      OptimizationDef arm64_optimizations[] = {
        OptDef(OptimizationPass::kInstructionSimplifierArm64),
```

```cpp
            OptDef(OptimizationPass::kSideEffectsAnalysis),
            OptDef(OptimizationPass::kGlobalValueNumbering, "GVN$after_arch"),
            OptDef(OptimizationPass::kScheduling)
        };
        return RunOptimizations(graph,
                                codegen,
                                dex_compilation_unit,
                                pass_observer,
                                handles,
                                arm64_optimizations);
    }
#endif
#ifdef ART_ENABLE_CODEGEN_mips
    case InstructionSet::kMips: {
        OptimizationDef mips_optimizations[] = {
            OptDef(OptimizationPass::kInstructionSimplifierMips),
            OptDef(OptimizationPass::kSideEffectsAnalysis),
            OptDef(OptimizationPass::kGlobalValueNumbering, "GVN$after_arch"),
            OptDef(OptimizationPass::kPcRelativeFixupsMips)
        };
        return RunOptimizations(graph,
                                codegen,
                                dex_compilation_unit,
                                pass_observer,
                                handles,
                                mips_optimizations);
    }
#endif
#ifdef ART_ENABLE_CODEGEN_mips64
    case InstructionSet::kMips64: {
        OptimizationDef mips64_optimizations[] = {
            OptDef(OptimizationPass::kSideEffectsAnalysis),
            OptDef(OptimizationPass::kGlobalValueNumbering, "GVN$after_arch")
        };
        return RunOptimizations(graph,
                                codegen,
                                dex_compilation_unit,
                                pass_observer,
                                handles,
                                mips64_optimizations);
    }
#endif
#ifdef ART_ENABLE_CODEGEN_x86
    case InstructionSet::kX86: {
        OptimizationDef x86_optimizations[] = {
            OptDef(OptimizationPass::kInstructionSimplifierX86),
```

```
          OptDef(OptimizationPass::kSideEffectsAnalysis),
          OptDef(OptimizationPass::kGlobalValueNumbering, "GVN$after_arch"),
          OptDef(OptimizationPass::kPcRelativeFixupsX86),
          OptDef(OptimizationPass::kX86MemoryOperandGeneration)
      };
      return RunOptimizations(graph,
                              codegen,
                              dex_compilation_unit,
                              pass_observer,
                              handles,
                              x86_optimizations);
    }
#endif
#ifdef ART_ENABLE_CODEGEN_x86_64
    case InstructionSet::kX86_64: {
      OptimizationDef x86_64_optimizations[] = {
          OptDef(OptimizationPass::kInstructionSimplifierX86_64),
          OptDef(OptimizationPass::kSideEffectsAnalysis),
          OptDef(OptimizationPass::kGlobalValueNumbering, "GVN$after_arch"),
          OptDef(OptimizationPass::kX86MemoryOperandGeneration)
      };
      return RunOptimizations(graph,
                              codegen,
                              dex_compilation_unit,
                              pass_observer,
                              handles,
                              x86_64_optimizations);
    }
#endif
    default:
      return false;
  }
}
```

从上述分析可以看出，最终的优化执行都是通过 7 个参数的 RunOptimizations() 函数去执行的，而 7 个参数的 RunOptimizations() 函数的参数是为 6 个参数的 RunOptimizations() 函数进行准备的，而 6 个参数的 RunOptimizations() 函数所需要的重要的一项参数就是具体优化 pass 的列表，这个准备工作可以在 5 个参数的 RunOptimizations() 函数中进行准备，也可以在 RunArchOptimizations() 函数中进行准备，不同的地方准备的是不同的优化列表。

本部分针对 ART IR 优化的主要流程进行了分析，关于 ART IR 优化的具体优化算法及其实现在本部分并没有深入介绍，这部分内容将在后续章节进行介绍。

3.5 寄存器分配

OptimizingCompiler 的寄存器分配，是紧跟着优化部分执行的环节，本部分将对其进行介绍。

寄存器分配是编译器优化领域中的一个常见环节，它主要为了解决程序中变量数量太多与 CPU 中寄存器数量有限之间的冲突，通过尽可能地将要使用的变量合理地放置在寄存器中，最终提升程序的执行效率。在这个过程中，通常会对变量的存活性进行分析，以便为活跃的变量分配寄存器。

寄存器分配环节的入口代码位于 art/compiler/optimizing/optimizing_compiler.cc 文件的 AllocateRegisters() 函数中，本部分将对其进行分析。本部分内容会涉及 HGraphDelegateVisitor 及 Accept() 函数的内容，可以预先阅读 4.2 节内容，这样能更加方便地理解本部分内容。

AllocateRegisters() 函数中进行了寄存器分配准备、SSA 存活性分析和寄存器分配等操作，代码如下：

```
//第 3 章/optimizing_compiler.cc
static void AllocateRegisters(HGraph* graph,
    CodeGenerator* codegen,
    PassObserver* pass_observer,
    RegisterAllocator::Strategy strategy,
    OptimizingCompilerStats* stats) {
  {
  PassScope scope(
      PrepareForRegisterAllocation::
          kPrepareForRegisterAllocationPassName,
      pass_observer);
  PrepareForRegisterAllocation(graph,
      codegen->GetCompilerOptions(), stats).Run();
  }
  //Use local allocator shared by SSA liveness analysis and register
  //allocator. (Register allocator creates new objects in the liveness
  //data.)
  ScopedArenaAllocator local_allocator(graph->GetArenaStack());
  SsaLivenessAnalysis liveness(graph, codegen, &local_allocator);
  {
    PassScope scope(SsaLivenessAnalysis::kLivenessPassName,
                pass_observer);
    liveness.Analyze();
  }
  {
    PassScope scope(RegisterAllocator::kRegisterAllocatorPassName,
```

```
                             pass_observer);
      std::unique_ptr<RegisterAllocator> register_allocator =
          RegisterAllocator::Create(&local_allocator, codegen, liveness,
                                    strategy);
      register_allocator->AllocateRegisters();
    }
}
```

寄存器分配准备、SSA 存活性分析和寄存器分配这 3 个环节在实现上分别对应着 PrepareForRegisterAllocation 类、SsaLivenessAnalysis 类和 RegisterAllocator 类，下面将对这 3 个部分进行分析。

3.5.1　PrepareForRegisterAllocation

PrepareForRegisterAllocation 类的定义和实现位于 art/compiler/optimizing/ 目录的 prepare_for_register_allocation.h 和 prepare_for_register_allocation.cc 中。

PrepareForRegisterAllocation 是一个 pass，它所对应的 pass 的名字为 prepare_for_register_allocation，和优化环节的 pass 有些类似，属于不同模块下的 pass。PrepareForRegisterAllocation 类继承于 HGraphDelegateVisitor 类，HGraphDelegateVisitor 类则继承于 HgraphVisitor 类，整个继承体系构成了一种根据需求遍历访问并且进行操作的框架。

关于 HGraphDelegateVisitor 类及其父类 HgraphVisitor 所构成的框架如何实现遍历访问和具体操作的内容，因为在优化部分已经涉及了很多，所以该部分内容放到优化实例的具体展开内容（4.2 节）进行介绍，在此不再展开。

PrepareForRegisterAllocation 类作为一个 pass，其核心函数是 Run() 函数，AllocateRegisters() 函数中也调用了 Run() 函数。Run() 函数的实现位于 prepare_for_register_allocation.cc 中，代码如下：

```
//第 3 章/prepare_for_register_allocation.cc
void PrepareForRegisterAllocation::Run() {
  //Order does not matter
  for (HBasicBlock* block : GetGraph()->GetReversePostOrder()) {
    //No need to visit the phis
    for (HInstructionIteratorHandleChanges
        inst_it(block->GetInstructions());
        !inst_it.Done(); inst_it.Advance()) {
      inst_it.Current()->Accept(this);
    }
  }
}
```

Run() 函数在这里遍历了除了 phi 指令之外的所有指令，并且对其指令调用了 Accept() 函数。有关Accept()函数的介绍也在HGraphDelegateVisitor相关部分进行介绍(4.2节)。最终,这里的Accept()函数会被转化为根据指令类型调用PrepareForRegisterAllocation类中的VisitXXXX()函数,如果指令没有其对应的VisitXXXX()函数或者其指令类型对应的VisitXXXX()函数,则它最终执行的是一个空操作,这是由HGraphDelegateVisitor类及其父类HgraphVisitor所设计的遍历架构解决的。

PrepareForRegisterAllocation类中实现了VisitCheckCast()函数、VisitInstanceOf()函数、VisitNullCheck()函数、VisitDivZeroCheck()函数、VisitBoundsCheck()函数、VisitBoundType()函数、VisitArraySet()函数、VisitClinitCheck()函数、VisitCondition()函数、VisitConstructorFence()函数、VisitInvokeStaticOrDirect()函数、VisitDeoptimize()函数和VisitTypeConversion()函数,也就是说除了这些函数所对应的指令,其他的指令的Accept()函数调用最终都是空操作。这些在PrepareForRegisterAllocation类中实现的VisitXXXX()函数就是真正要为寄存器分配所做的准备工作。

以PrepareForRegisterAllocation的VisitDivZeroCheck()函数为例,它实现了对DivZeroCheck指令Accept()函数的具体内容,DivZeroCheck在分类上属于Instruction,所以VisitDivZeroCheck()函数属于直接针对具体指令,而不是针对指令分类。具体指令及其分类的对应关系,可以在art/compiler/optimizing/nodes.h中的相关宏中查到。

VisitDivZeroCheck()函数的具体实现位于prepare_for_register_allocation.cc中,代码如下:

```
//第3章/prepare_for_register_allocation.cc
void PrepareForRegisterAllocation::VisitDivZeroCheck(
      HDivZeroCheck * check) {
  check->ReplaceWith(check->InputAt(0));
}
```

其中对DivZeroCheck()函数直接进行了替换。需要说明的是,PrepareForRegisterAllocation类作为寄存器分配前的准备工作,其所做的主要工作是:It changes uses of null checks and bounds checks to the original objects, to avoid creating a live range for these checks.,改变null检查和边界检查来避免为这些检查创建live range。

3.5.2 SsaLivenessAnalysis

SsaLivenessAnalysis类对应的是SSA存活性分析,SSA存活性分析主要是为了后续的真正的寄存器分配做信息准备,本部分将对SsaLivenessAnalysis类的相关实现进行分析。

SsaLivenessAnalysis的定义和实现位于art/compiler/optimizing/目录下的ssa_liveness_analysis.h和ssa_liveness_analysis.cc中。

AllocateRegisters()函数中对SsaLivenessAnalysis类的使用,就是创建对象liveness

之后,通过对象调用其 Analyze() 函数。SsaLivenessAnalysis::Analyze() 函数的实现位于 ssa_liveness_analysis.cc 中,代码如下:

```
//第 3 章/ssa_liveness_analysis.cc
void SsaLivenessAnalysis::Analyze() {
  //Compute the linear order directly in the graph's data structure
  //(there are no more following graph mutations)
  LinearizeGraph(graph_, &graph_->linear_order_);

  //Liveness analysis
  NumberInstructions();
  ComputeLiveness();
}
```

其中,LinearizeGraph() 函数负责将图中的结构线性化;NumberInstructions() 函数给每个指令一个 SSA 编号;ComputeLiveness() 函数负责计算指令的存活区间,主要包括 live_in、live_out 和 kill 这几个集合。这些信息都是为了后续进行实际的寄存器分配做准备的。

3.5.3 RegisterAllocator

寄存器分配所对应的实际操作主要位于与 RegisterAllocator 类相关的处理中,这是真正进行寄存器分配的环节,前面的环节都是为寄存器分配准备的相关信息。

RegisterAllocator 类的声明和实现位于 art/compiler/optimizing/register_allocator.h 和 register_allocator.cc 中。从 register_allocator.h 中 RegisterAllocator 类的 Strategy 可以看到,ART 的寄存器分配有两种策略:线性扫描和图着色。Strategy 的具体代码如下:

```
//第 3 章/register_allocator.h
  enum Strategy {
    kRegisterAllocatorLinearScan,
    kRegisterAllocatorGraphColor
  };
```

并且,在 RegisterAllocator 类中,还有一个成员变量 kRegisterAllocatorDefault 对默认策略进行了设置,将其设置为 kRegisterAllocatorLinearScan,也就是线性扫描策略。具体的代码如下:

```
static constexpr Strategy kRegisterAllocatorDefault =
        kRegisterAllocatorLinearScan;
```

从这里已经可以确定,ART 的寄存器分配采用线性扫描和图着色,并且线性扫描是默认的分配策略。在本节刚开始的位置,其中有关实际进行寄存器分配的操作调用了

RegisterAllocator::Create()函数,并根据其返回的指针调用了AllocateRegisters()函数。

RegisterAllocator::Create()函数的实现位于 register_allocator.cc 中,它会根据所采用的线性扫描策略或者图着色策略构建对应的 RegisterAllocatorLinearScan 对象或者 RegisterAllocatorGraphColor 对象,然后返回指向该对象的智能指针。具体的代码如下：

```
//第 3 章/register_allocator.cc
std::unique_ptr<RegisterAllocator>
RegisterAllocator::Create(ScopedArenaAllocator* allocator,
                          CodeGenerator* codegen,
                          const SsaLivenessAnalysis& analysis,
                          Strategy strategy) {
  switch (strategy) {
    case kRegisterAllocatorLinearScan:
      return std::unique_ptr<RegisterAllocator>(
          new (allocator) RegisterAllocatorLinearScan(allocator, codegen,
                                                     analysis));
    case kRegisterAllocatorGraphColor:
      return std::unique_ptr<RegisterAllocator>(
          new (allocator) RegisterAllocatorGraphColor(allocator, codegen,
                                                     analysis));
    default:
      LOG(FATAL) << "Invalid register allocation strategy: " << strategy;
      UNREACHABLE();
  }
}
```

RegisterAllocator 类的 AllocateRegisters()函数本身是一个纯虚函数,所以最终对于 AllocateRegisters()函数的调用,实际上调用了 RegisterAllocatorLinearScan 或者 RegisterAllocatorGraphColor 的 AllocateRegisters()函数。

RegisterAllocatorLinearScan 类和 RegisterAllocatorGraphColor 类都有单独的实现文件, RegisterAllocatorLinearScan 类的声明和实现位于 art/compiler/optimizing/register_allocator_liner_scan. h 和 register_allocator_liner_scan. cc 中。RegisterAllocatorGraphColor 类的声明和实现位于 art/compiler/optimizing/register_allocator_graph_color. h 和 register_allocator_graph_color. cc 中。这两个类在自身对应的 cc 文件中,都由具体的 AllocateRegisters()函数实现。由于线性扫描和图着色都是经典的寄存器分配策略,所以这里对具体的实现就不展开分析了。

至此,本部分完成了对寄存器分配内容的介绍,其中只有第 3 部分实际进行了寄存器的分配,前面的部分都是为最终的寄存器分配做准备的。ART 默认使用的是线性扫描的寄存器分配策略,备用的分配策略是图着色策略,这都是目前经典的寄存器分配策略。

3.6 代码生成

OptimizingCompiler 的代码生成部分对应的具体实现是 art/compiler/optimizing 中的 CodeGenerator 类。本部分内容将对其进行分析。

OptimizingCompiler 的代码生成部分发生在宏观 OptimizingCompiler 进行优化和寄存器分配的环节之后，在具体代码实现上，上层是在 art/compiler/optimizing/optimizing_compiler.cc 的 OptimizingCompiler::Compile() 函数中通过 codegen.reset() 函数的形式体现。因为 codegen 是声明为 std::unique_ptr<CodeGenerator>的智能指针，所以使用了 reset 更新其内容，而 codegen.reset() 函数的输入是 TryCompileIntrinsic() 函数或 TryCompile() 函数的返回值。TryCompileIntrinsic() 函数和 TryCompile() 函数都通过 CodeGenerator::Create() 函数为 std::unique_ptr<CodeGenerator> codegen 赋初值，通过 codegen->GetAssembler()->cfi().SetEnabled() 进行 Assembler 相关操作，之后通过 codegen->Compile() 函数实现代码生成的主要内容，最终通过 return codegen.release() 函数实现对 OptimizingCompiler::Compile() 函数中的 codegen 进行赋值。以 TryCompile() 函数为例，代码如下：

```
//第 3 章/optimizing_compiler.cc
CodeGenerator* OptimizingCompiler::TryCompile(
    ArenaAllocator* allocator,
    ArenaStack* arena_stack,
    CodeVectorAllocator* code_allocator,
    const DexCompilationUnit& dex_compilation_unit,
    ArtMethod* method, bool baseline, bool osr,
    VariableSizedHandleScope* handles) const {
…

  std::unique_ptr<CodeGenerator> codegen(
      CodeGenerator::Create(graph, compiler_options,
                            compilation_stats_.get()));
  if (codegen.get() == nullptr) {
    MaybeRecordStat(compilation_stats_.get(),
                    MethodCompilationStat::kNotCompiledNoCodegen);
    return nullptr;
  }
  codegen->GetAssembler()->cfi().SetEnabled(
      compiler_options.GenerateAnyDebugInfo());
…
  codegen->Compile(code_allocator);
  pass_observer.DumpDisassembly();

  MaybeRecordStat(compilation_stats_.get(),
```

```
                    MethodCompilationStat::kCompiledBytecode);
  return codegen.release();
}
```

在这里,最终返回了 codegen.release()函数,这个就是要给 OptimizingCompiler::Compile()函数中 codegen.reset()函数所使用的输入,这时候代码生成部分其实已经完成了,所以 CodeGenerator 类需要重点关注的几个成员函数主要有 Create()函数和 GetAssembler()函数之后的 Assembler 相关操作和 Compile()函数。

CodeGenerator 类的实现主要位于 art/compiler/optimizing 中的 code_generator.h 和 code_generator.cc 中。其中,Create()函数会根据代码生成的具体硬件目标平台,去调用对应类的构造函数,代码如下:

```
//第3章/code_generator.cc
std::unique_ptr<CodeGenerator> CodeGenerator::Create(
HGraph* graph, const CompilerOptions& compiler_options,
    OptimizingCompilerStats* stats) {
  ArenaAllocator* allocator = graph->GetAllocator();
  switch (compiler_options.GetInstructionSet()) {
#ifdef ART_ENABLE_CODEGEN_arm
    case InstructionSet::kArm:
    case InstructionSet::kThumb2: {
      return std::unique_ptr<CodeGenerator>(
          new (allocator) arm::CodeGeneratorARMVIXL(graph, compiler_options, stats));
    }
#endif
#ifdef ART_ENABLE_CODEGEN_arm64
    case InstructionSet::kArm64: {
      return std::unique_ptr<CodeGenerator>(
          new (allocator) arm64::CodeGeneratorARM64(graph, compiler_options, stats));
    }
#endif
#ifdef ART_ENABLE_CODEGEN_mips
    case InstructionSet::kMips: {
      return std::unique_ptr<CodeGenerator>(
          new (allocator) mips::CodeGeneratorMIPS(graph, compiler_options, stats));
    }
#endif
#ifdef ART_ENABLE_CODEGEN_mips64
    case InstructionSet::kMips64: {
      return std::unique_ptr<CodeGenerator>(
          new (allocator) mips64::CodeGeneratorMIPS64(graph, compiler_options, stats));
    }
#endif
#ifdef ART_ENABLE_CODEGEN_x86
```

```
      case InstructionSet::kX86: {
        return std::unique_ptr<CodeGenerator>(
            new (allocator) x86::CodeGeneratorX86(graph, compiler_options, stats));
      }
#endif
#ifdef ART_ENABLE_CODEGEN_x86_64
      case InstructionSet::kX86_64: {
        return std::unique_ptr<CodeGenerator>(
            new (allocator) x86_64::CodeGeneratorX86_64(graph, compiler_options, stats));
      }
#endif
      default:
        return nullptr;
    }
}
```

从这里可以看到,根据 arm、arm64、mips、mips64、x86 和 x86_64 这几个不同的硬件平台,会调用每个平台对应的 CodeGeneratorXXX 类的构造函数。针对每个具体的目标平台,都有一个 CodeGenerator 类的子类对应实现相关内容,并且有对应的.h 和.cc 文件专门用于实现存储。以 ARM64 为例,class CodeGeneratorARM64 : public CodeGenerator,相关定义和实现位于 art/compiler/optimizing 中的 code_generator_arm64.h 和 code_generator_arm64.cc。

所以,从 CodeGenerator 类的 Create()函数开始,包括之后 GetAssembler()函数的 Assembler 相关操作和 Compile()函数都要最终落在硬件平台相关的子类上,以 ARM64 为例,GetAssembler()函数之后 Assembler 相关操作和 Compile()函数最终要看 CodeGeneratorARM64 中的实现。

CodeGeneratorARM64 中的 GetAssembler()函数实现,代码如下:

```
//第 3 章/code_generator_arm64.h
Arm64Assembler* GetAssembler() override { return &assembler_; }
  const Arm64Assembler& GetAssembler() const override {
      return assembler_;
}
```

其中,assembler_的声明是 Arm64Assembler assembler_,Arm64Assembler 的实现位于 art/compiler/utils/arm64/中的 assembler_arm64.h 和 assembler_arm64.cc。它是 Assembler 的子类,声明如下:

```
class Arm64Assembler final : public Assembler {
```

每个硬件平台,除了有自己对应的 CodeGenerator 子类,还有自己对应的 Assembler 子类。Assembler 类的具体实现位于 art/compiler/utils/中的 assembler.h 和 assembler.cc 中,而各个硬件平台所对应的 Assembler 子类都位于 art/compiler/utils/下各自对立的目录,分别为 arm、arm64、mips、mips64、x86 和 x86_64。

这时候,codegen －＞ GetAssembler( ) －＞ cfi( ). SetEnabled( )的操作就变成了对 Arm64Assembler 的 cfi( ). SetEnabled( )函数进行操作。Arm64Assembler 并未直接定义 cfi( )函数,此时会调用其父类 Assembler 有关 cfi( )函数的定义,代码如下:

```
DebugFrameOpCodeWriterForAssembler& cfi() { return cfi_; }
```

这里的 cfi_的定义如下:

```
DebugFrameOpCodeWriterForAssembler cfi_;
```

DebugFrameOpCodeWriterForAssembler 的定义也位于 art/compiler/utils/assembler.h 中,然后对 SetEnabled( )函数进行了操作。

CodeGeneratorARM64 类并没有 Compile( )函数,这里从其父类 CodeGenerator 中继承了 Compile( )函数,代码如下:

```
//第3章/code_generator.cc
void CodeGenerator::Compile(CodeAllocator* allocator) {
    InitializeCodeGenerationData();

    //The register allocator already called InitializeCodeGeneration
    //where the frame size has been computed
    DCHECK(block_order_ != nullptr);
    Initialize();

    HGraphVisitor* instruction_visitor = GetInstructionVisitor();
    DCHECK_EQ(current_block_index_, 0u);

    GetStackMapStream()->BeginMethod(HasEmptyFrame() ? 0 : frame_size_,
                                     core_spill_mask_,
                                     fpu_spill_mask_,
                                     GetGraph()->GetNumberOfVRegs());

    size_t frame_start = GetAssembler()->CodeSize();
    GenerateFrameEntry();
    DCHECK_EQ(GetAssembler()->cfi().GetCurrentCFAOffset(), static_cast<int>(frame_size_));
    if (disasm_info_ != nullptr) {
        disasm_info_->SetFrameEntryInterval(frame_start,
```

```cpp
                                            GetAssembler()->CodeSize());
  }

  for (size_t e = block_order_->size(); current_block_index_ < e;
       ++current_block_index_) {
    HBasicBlock* block = (*block_order_)[current_block_index_];
    //Don't generate code for an empty block. Its predecessors will
    //branch to its successor directly. Also, the label of that block
    //will not be emitted, so this helps catch errors where we reference
    //that label
    if (block->IsSingleJump()) continue;
    Bind(block);
    //This ensures that we have correct native line mapping for all
    //native instructions. It is necessary to make stepping over a
    //statement work. Otherwise, any initial instructions (e.g. moves)
    //would be assumed to be the start of next statement
    MaybeRecordNativeDebugInfo(/* instruction= */ nullptr,
                               block->GetDexPc());
    for (HInstructionIterator it(block->GetInstructions()); !it.Done();
         it.Advance()) {
      HInstruction* current = it.Current();
      if (current->HasEnvironment()) {
        //Create stackmap for HNativeDebugInfo or any instruction which
        //calls native code. Note that we need correct mapping for the
        //native PC of the call instruction, so the Runtime's stackmap is
        //not sufficient since it is at PC after the call
        MaybeRecordNativeDebugInfo(current, block->GetDexPc());
      }
      DisassemblyScope disassembly_scope(current, *this);
      DCHECK(CheckTypeConsistency(current));
      current->Accept(instruction_visitor);
    }
  }

  GenerateSlowPaths();

  //Emit catch stack maps at the end of the stack map stream as expected
  //by the Runtime exception handler
  if (graph_->HasTryCatch()) {
    RecordCatchBlockInfo();
  }

  //Finalize instructions in assembler
  Finalize(allocator);

  GetStackMapStream()->EndMethod();
}
```

这样就完成了 ARM64 平台的代码生成环节。其他的硬件平台也有类似的执行流程，可以根据硬件平台的实际需要去阅读具体的代码。这里需要说明的是，代码生成环节完全依赖于硬件平台，每个硬件平台都有不同的需求与实现，对于不同的硬件平台需要完全区分。

本部分对 OptimizingCompiler 的代码生成环节进行了介绍，由于代码生成环节严格依赖于硬件平台，所以本章在介绍时选择 ARM64 平台的代码作为具体的例子进行了展示，其他平台的具体实现可以参照对应的代码。

3.7 OptimizingCompiler 总结

OptimizingCompiler 在 ART 中既承担着 dex 到 oat 的编译过程，也承担着 Runtime 中的 JIT 编译过程。这两部分对 ART 来讲都很重要，也就是说整个 ART 体系之下，所进行的核心编译工作都是由 OptimizingCompiler 完成的，如图 3.2 所示。

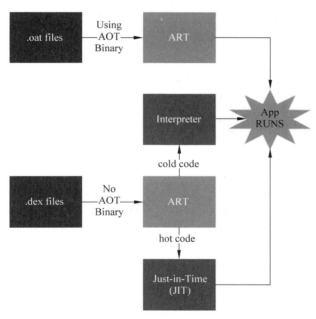

图 3.2 ART 体系示意图

（图源：https://source.android.com/devices/tech/dalvik/jit-compiler）

其中，从 dex 到 oat 的编译过程是在 dex2oat 工具中被调用的，其内部调用关系如图 3.3 所示。

图 3.3 右上方方框内是 OptimizingCompiler 类的两个成员函数，这两个函数分别对普通方法和 JNI 方法进行编译，是 OptimizingCompiler 编译的主要内容。

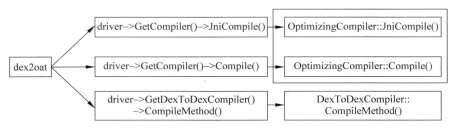

图 3.3　dex2oat 内部调用示意图

ART 内部的 Runtime 在运行的时候，会根据情况调用 JIT 对待执行方法进行编译，这个过程需要调用 JitCompiler 类的 CompileMethod() 函数，最终调用的是 OptimizingCompiler 中 OptimizingCompiler 类的 JitCompile() 函数，如图 3.4 右上方方框中所示。

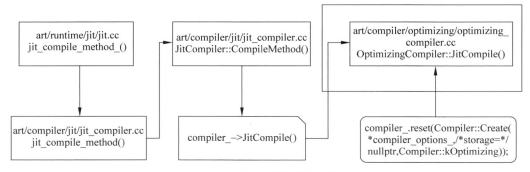

图 3.4　JIT 内部调用示意图

其中，compiler_.reset() 函数对 compiler_ 的值进行了设置，通过 Compiler::Create() 函数创建了 OptimizingCompiler 类的对象，所以 compiler_—>JitCompile() 的调用最终被转化为 OptimizingCompiler 类的 JitCompile() 函数。

上述内容总结了 OptimizingCompiler 在整个 ART 中的任务，除了在 ART 的解释执行流程中用不到 OptimizingCompiler 之外，其余的 dex2oat 流程和 JIT 流程都需要用到 OptimizingCompiler 进行编译，所以 OptimizingCompiler 的相关内容要进行认真理解。

3.8　小结

本章首先对宏观 OptimizingCompiler 中 JNI 处理和普通方法编译进行了介绍，之后对其中都涉及的 HGraph 的构建、优化、寄存器分配和代码生成环节进行了介绍，最后对 OptimizingCompiler 进行了介绍，将前面的一些流程进行了整体上的梳理。OptimizingCompiler 对于 dex2oat 和 ART 的 JIT 部分都是非常重要的核心内容，需要深入理解。

第 4 章 OptimizingCompiler 优化算法分析

OptimizingCompiler 中有针对 ART IR 进行优化的环节,这个环节通过十几个优化算法的顺序执行实现了对 ART IR 的优化,本部分内容将对这些优化算法进行具体分析。

4.1 优化算法框架

在 3.4 节介绍 OptimizingCompiler 的时候提到过它的优化环节。OptimizingCompiler 有十几种优化算法,每种优化算法都由具体的类去实现,在管理和实现上都在同一个框架下,本节将对这个框架进行简要介绍。

在 3.4 节介绍 ART IR 的优化环节的时候提到过在 5 个参数版本的 OptimizingCompiler::RunOptimizations()函数(art/compiler/optimizing/optimizing_compiler.cc)中,通过数组的形式准备所有要执行的 pass,代码如下:

```
//第 4 章/ optimizing_compiler.cc
OptimizationDef optimizations[] = {
    //Initial optimizations
    OptDef(OptimizationPass::kConstantFolding),
    OptDef(OptimizationPass::kInstructionSimplifier),
    OptDef(OptimizationPass::kDeadCodeElimination,
        "dead_code_elimination $ initial"),
    //Inlining
    OptDef(OptimizationPass::kInliner),
    //Simplification (only if inlining occurred)
    OptDef(OptimizationPass::kConstantFolding,
        "constant_folding $ after_inlining",
        OptimizationPass::kInliner),
    OptDef(OptimizationPass::kInstructionSimplifier,
        "instruction_simplifier $ after_inlining",
        OptimizationPass::kInliner),
    OptDef(OptimizationPass::kDeadCodeElimination,
        "dead_code_elimination $ after_inlining",
        OptimizationPass::kInliner),
```

```
    //GVN
    OptDef(OptimizationPass::kSideEffectsAnalysis,
           "side_effects$before_gvn"),
    OptDef(OptimizationPass::kGlobalValueNumbering),
    //Simplification (TODO: only if GVN occurred)
    OptDef(OptimizationPass::kSelectGenerator),
    OptDef(OptimizationPass::kConstantFolding,
           "constant_folding$after_gvn"),
    OptDef(OptimizationPass::kInstructionSimplifier,
           "instruction_simplifier$after_gvn"),
    OptDef(OptimizationPass::kDeadCodeElimination,
           "dead_code_elimination$after_gvn"),
    //High-level optimizations
    OptDef(OptimizationPass::kSideEffectsAnalysis,
           "side_effects$before_licm"),
    OptDef(OptimizationPass::kInvariantCodeMotion),
    OptDef(OptimizationPass::kInductionVarAnalysis),
    OptDef(OptimizationPass::kBoundsCheckElimination),
    OptDef(OptimizationPass::kLoopOptimization),
    //Simplification
    OptDef(OptimizationPass::kConstantFolding,
           "constant_folding$after_bce"),
    OptDef(OptimizationPass::kInstructionSimplifier,
           "instruction_simplifier$after_bce"),
    //Other high-level optimizations
    OptDef(OptimizationPass::kSideEffectsAnalysis,
           "side_effects$before_lse"),
    OptDef(OptimizationPass::kLoadStoreAnalysis),
    OptDef(OptimizationPass::kLoadStoreElimination),
    OptDef(OptimizationPass::kCHAGuardOptimization),
    OptDef(OptimizationPass::kDeadCodeElimination,
           "dead_code_elimination$final"),
    OptDef(OptimizationPass::kCodeSinking),
    //The codegen has a few assumptions that only the instruction
    //simplifier can satisfy. For example, the code generator does not
    //expect to see a HTypeConversion from a type to the same type
    OptDef(OptimizationPass::kInstructionSimplifier,
           "instruction_simplifier$before_codegen"),
    //Eliminate constructor fences after code sinking to avoid
    //complicated sinking logic to split a fence with many inputs
    OptDef(OptimizationPass::kConstructorFenceRedundancyElimination)
};
```

optimizations 数组中的元素都是 OptimizationDef 类型的，数组内的元素都是通过 OptDef(OptimizationPass::kConstantFolding)这种形式来定义的。其中，OptDef()函数是一个内联函数，它通过输入的参数，构建出一个 OptimizationDef 结构体对象，它的实现位于

art/compiler/optimizing/optimization.h 中,代码如下:

```
//第 4 章/ optimization.h
//Helper method for optimization definition array entries
inline OptimizationDef OptDef(
    OptimizationPass pass, const char * pass_name = nullptr,
    OptimizationPass depends_on = OptimizationPass::kNone) {
  return OptimizationDef(pass, pass_name, depends_on);
}
```

OptimizationDef 结构体的定义也位于 art/compiler/optimizing/optimization.h 中,代码如下:

```
//第 4 章/ optimization.h
//Optimization definition consisting of an optimization pass
//an optional alternative name (nullptr denotes default), and
//an optional pass dependence (kNone denotes no dependence)
struct OptimizationDef {
  OptimizationDef(OptimizationPass p, const char * pn, OptimizationPass d)
      : pass(p), pass_name(pn), depends_on(d) {}
  OptimizationPass pass;
  const char * pass_name;
  OptimizationPass depends_on;
};
```

通过上述代码,可以清楚地看出,要构建一个 OptimizationDef 结构体对象,需要一个 pass,并且有其对应的字符串形式的 pass_name,同时还有一个 pass 所依赖的 pass 变量 depends_on。pass_name 在默认情况下是空的,depends_on 在默认情况下为 OptimizationPass::kNone,这相当于在默认情况下,pass 是不依赖于其他 pass 的,如果有依赖,则需要为 depends_on 传递新值。

在 optimizations 数组中,有 kConstantFolding、kInstructionSimplifier、kDeadCodeElimination、kSideEffectsAnalysis、kGlobalValueNumbering、kSelectGenerator、kInvariantCodeMotion、kInductionVarAnalysis、kBoundsCheckElimination、kLoopOptimization、kLoadStoreAnalysis、kLoadStoreElimination、kCHAGuardOptimization、kCodeSinking 和 kConstructorFenceRedundancyElimination 共 15 个 pass,其中 kInstructionSimplifier 和 kDeadCodeElimination 等 pass 还被执行了多次。这些 pass 都是 OptimizationPass 类型的,OptimizationPass 的定义位于 art/compiler/optimizing/optimization.h 中,代码如下:

```
//第 4 章/ optimization.h
//Optimization passes that can be constructed by the helper method
//below. An enum field is preferred over a string lookup at places where
```

```
//performance matters. TODO: generate this table and lookup methods
//below automatically
enum class OptimizationPass {
  kBoundsCheckElimination,
  kCHAGuardOptimization,
  kCodeSinking,
  kConstantFolding,
  kConstructorFenceRedundancyElimination,
  kDeadCodeElimination,
  kGlobalValueNumbering,
  kInductionVarAnalysis,
  kInliner,
  kInstructionSimplifier,
  kInvariantCodeMotion,
  kLoadStoreAnalysis,
  kLoadStoreElimination,
  kLoopOptimization,
  kScheduling,
  kSelectGenerator,
  kSideEffectsAnalysis,
#ifdef ART_ENABLE_CODEGEN_arm
  kInstructionSimplifierArm,
#endif
#ifdef ART_ENABLE_CODEGEN_arm64
  kInstructionSimplifierArm64,
#endif
#ifdef ART_ENABLE_CODEGEN_mips
  kPcRelativeFixupsMips,
  kInstructionSimplifierMips,
#endif
#ifdef ART_ENABLE_CODEGEN_x86
  kPcRelativeFixupsX86,
  kInstructionSimplifierX86,
#endif
#ifdef ART_ENABLE_CODEGEN_x86_64
  kInstructionSimplifierX86_64,
#endif
#if defined(ART_ENABLE_CODEGEN_x86) || defined(ART_ENABLE_CODEGEN_x86_64)
  kX86MemoryOperandGeneration,
#endif
  kNone,
  kLast = kNone
};
```

这些 OptimizationPass 类型的 pass，会在 7 个参数版本的 RunOptimizations()函数（art/compiler/optimizing/optimizing_compiler.cc）中被转换为 ArenaVector＜HOptimization＊＞类

型，这种类型中的每一项都是一个 HOptimization * 类型，它是真正要执行的 pass 实体。这个转换过程是在 ConstructOptimizations() 函数中实现的，它位于 art/compiler/optimizing/optimization.cc 中，代码如下：

```cpp
//第 4 章/ optimization.cc
ArenaVector<HOptimization*> ConstructOptimizations(
    const OptimizationDef definitions[],
    size_t length,
    ArenaAllocator* allocator,
    HGraph* graph,
    OptimizingCompilerStats* stats,
    CodeGenerator* codegen,
    const DexCompilationUnit& dex_compilation_unit,
    VariableSizedHandleScope* handles) {
  ArenaVector<HOptimization*> optimizations(allocator->Adapter());

  //Some optimizations require SideEffectsAnalysis or
  //HInductionVarAnalysis instances. This method uses the nearest
  //instance preceeding it in the pass name list or fails fatally if no such analysis can be found
  SideEffectsAnalysis* most_recent_side_effects = nullptr;
  HInductionVarAnalysis* most_recent_induction = nullptr;
  LoadStoreAnalysis* most_recent_lsa = nullptr;

  //Loop over the requested optimizations
  for (size_t i = 0; i < length; i++) {
    OptimizationPass pass = definitions[i].pass;
    const char* alt_name = definitions[i].pass_name;
    const char* pass_name = alt_name != nullptr
        ? alt_name
        : OptimizationPassName(pass);
    HOptimization* opt = nullptr;

    switch (pass) {
      //
      //Analysis passes (kept in most recent for subsequent passes)
      //
      case OptimizationPass::kSideEffectsAnalysis:
        opt = most_recent_side_effects =
            new (allocator) SideEffectsAnalysis(graph, pass_name);
        break;
      case OptimizationPass::kInductionVarAnalysis:
        opt = most_recent_induction =
            new (allocator) HInductionVarAnalysis(graph, pass_name);
        break;
      case OptimizationPass::kLoadStoreAnalysis:
```

```cpp
      opt = most_recent_lsa =
          new (allocator) LoadStoreAnalysis(graph, pass_name);
      break;
    //
    //Passes that need prior analysis
    //
    case OptimizationPass::kGlobalValueNumbering:
      CHECK(most_recent_side_effects != nullptr);
      opt = new (allocator) GVNOptimization(graph,
          *most_recent_side_effects, pass_name);
      break;
    case OptimizationPass::kInvariantCodeMotion:
      CHECK(most_recent_side_effects != nullptr);
      opt = new (allocator) LICM(graph,
          *most_recent_side_effects, stats, pass_name);
      break;
    case OptimizationPass::kLoopOptimization:
      CHECK(most_recent_induction != nullptr);
      opt = new (allocator) HLoopOptimization(graph,
          &codegen->GetCompilerOptions(),
          most_recent_induction, stats, pass_name);
      break;
    case OptimizationPass::kBoundsCheckElimination:
      CHECK(most_recent_side_effects != nullptr &&
          most_recent_induction != nullptr);
      opt = new (allocator) BoundsCheckElimination(graph,
          *most_recent_side_effects, most_recent_induction, pass_name);
      break;
    case OptimizationPass::kLoadStoreElimination:
      CHECK(most_recent_side_effects != nullptr &&
          most_recent_induction != nullptr);
      opt = new (allocator) LoadStoreElimination(graph,
          *most_recent_side_effects, *most_recent_lsa, stats, pass_name);
      break;
    //
    //Regular passes
    //
    case OptimizationPass::kConstantFolding:
      opt = new (allocator) HConstantFolding(graph, pass_name);
      break;
    case OptimizationPass::kDeadCodeElimination:
      opt = new (allocator) HDeadCodeElimination(graph, stats, pass_name);
      break;
    case OptimizationPass::kInliner: {
      CodeItemDataAccessor accessor(*dex_compilation_unit.GetDexFile(),
                                    dex_compilation_unit.GetCodeItem());
```

```cpp
      opt = new (allocator) HInliner(graph, graph, codegen,
                                      dex_compilation_unit,
                                      dex_compilation_unit,
                                      handles, stats,
                                      accessor.RegistersSize(),
                                      0, nullptr, 0, pass_name);
      break;
    }
    case OptimizationPass::kSelectGenerator:
      opt = new (allocator) HSelectGenerator(graph, handles, stats,
                                              pass_name);
      break;
    case OptimizationPass::kInstructionSimplifier:
      opt = new (allocator) InstructionSimplifier(graph, codegen, stats,
                                                   pass_name);
      break;
    case OptimizationPass::kCHAGuardOptimization:
      opt = new (allocator) CHAGuardOptimization(graph, pass_name);
      break;
    case OptimizationPass::kCodeSinking:
      opt = new (allocator) CodeSinking(graph, stats, pass_name);
      break;
    case OptimizationPass::kConstructorFenceRedundancyElimination:
      opt = new (allocator) ConstructorFenceRedundancyElimination(graph,
          stats, pass_name);
      break;
    case OptimizationPass::kScheduling:
      opt = new (allocator) HInstructionScheduling(graph,
          codegen->GetCompilerOptions().GetInstructionSet(), codegen, pass_name);
      break;
    //
    //Arch-specific passes
    //
#ifdef ART_ENABLE_CODEGEN_arm
    case OptimizationPass::kInstructionSimplifierArm:
      DCHECK(alt_name == nullptr) << "arch-specific pass does not
          support alternative name";
      opt = new (allocator) arm::InstructionSimplifierArm(graph, stats);
      break;
#endif
#ifdef ART_ENABLE_CODEGEN_arm64
    case OptimizationPass::kInstructionSimplifierArm64:
      DCHECK(alt_name == nullptr) << "arch-specific pass does not support alternative name";
      opt = new (allocator) arm64::InstructionSimplifierArm64(graph,
                                                               stats);
      break;
```

```cpp
# endif
# ifdef ART_ENABLE_CODEGEN_mips
        case OptimizationPass::kPcRelativeFixupsMips:
          DCHECK(alt_name == nullptr) << "arch-specific pass does not support alternative name";
          opt = new (allocator) mips::PcRelativeFixups(graph, codegen,
                                                      stats);
          break;
        case OptimizationPass::kInstructionSimplifierMips:
          DCHECK(alt_name == nullptr) << "arch-specific pass does not support alternative name";
          opt = new (allocator) mips::InstructionSimplifierMips(graph,
              codegen, stats);
          break;
# endif
# ifdef ART_ENABLE_CODEGEN_x86
        case OptimizationPass::kPcRelativeFixupsX86:
          DCHECK(alt_name == nullptr) << "arch-specific pass does not support alternative name";
          opt = new (allocator) x86::PcRelativeFixups(graph, codegen,
                                                     stats);
          break;
        case OptimizationPass::kX86MemoryOperandGeneration:
          DCHECK(alt_name == nullptr) << "arch-specific pass does not support alternative name";
          opt = new (allocator) x86::X86MemoryOperandGeneration(graph,
              codegen, stats);
          break;
        case OptimizationPass::kInstructionSimplifierX86:
         opt = new (allocator) x86::InstructionSimplifierX86(graph, codegen,
                                                             stats);
         break;
# endif
# ifdef ART_ENABLE_CODEGEN_x86_64
        case OptimizationPass::kInstructionSimplifierX86_64:
          opt = new (allocator) x86_64::InstructionSimplifierX86_64(graph,
              codegen, stats);
          break;
# endif
        case OptimizationPass::kNone:
          LOG(FATAL) << "kNone does not represent an actual pass";
          UNREACHABLE();
    } //switch

    //Add each next optimization to result vector
    CHECK(opt != nullptr);
    DCHECK_STREQ(pass_name, opt->GetPassName()); //sanity
    optimizations.push_back(opt);
  }

  return optimizations;
}
```

在这里，通过对 OptimizationPass 类型的 pass 采用 switch-case 结构进行选择，根据 pass 的具体值使用 graph、pass_name、stats 等参数构建具体的优化类，这些具体的优化类都是 HOptimization 的子类。在这里需要指出的是，并不是所有 OptimizationPass 类型的 pass 都有对应的 pass_name，对于没有 pass_name 的情况，会根据 OptimizationPass 类型的 pass 值通过 OptimizationPassName 来取 pass_name，代码如下：

```
const char * pass_name = alt_name != nullptr ? alt_name
    : OptimizationPassName(pass);
```

OptimizationPassName( ) 函数的实现位于 art/compiler/optimizing/optimization.cc 中，代码较为简单，通过 switch-case 结构针对 pass 返回 pass_name。这里所返回的 pass_name 是由每个 pass 所对应的实现类的内部所给出的。

上文提到具体的优化类都是 HOptimization 的子类，之前已经统计过与架构无关的 15 个 pass 都有自己对应的优化类，这些 HOptimization 的子类都有一个自己对应的头文件、cc 文件和一个测试文件。例如，kConstantFolding pass 所对应的优化类为 HConstantFolding，它继承于 HOptimization，它的定义和实现位于其对应的 art/compiler/optimizing/constant_floding.h 和 art/compiler/optimizing/constant_floding.cc 中，并且有一个对应的测试文件 art/compiler/optimizing/constant_floding_test.cc。其他的具体优化类也是类似的情况，这些内容会在后续对具体优化类介绍的时候进行分析。

优化类中，除了与架构无关的优化类，还有一批与架构相关的优化类，这种优化类针对具体目标平台进行了优化。这些优化类和架构无关的优化类的实现相同，都有对应的优化类，优化类都有自己对应的 .h 和 .cc 文件，只是没有对应的测试文件。例如，kInstructionSimplifierX86_64 pass 对应的优化类为 InstructionSimplifierX86_64，InstructionSimplifierX86_64 的定义和实现位于对应的 art/compiler/optimizing/instruction_simplifier_x86_64.h 和 art/compiler/optimizing/instruction_simplifier_x86_64.cc 中。

这里所涉及的所有的优化类都是 HOptimization 的子类，部分优化类会依赖于其他优化类，优化类的具体的执行内容需要重载 Run( ) 函数去实现。后续会在具体的优化类的分析过程中进行介绍。

4.2 常量折叠

常量折叠优化指的是将表达式中的常量计算在编译的时候就进行求值，这样可以提高运行时的性能，并且可以缩减代码的大小。本部分将对 OptimizingCompiler 优化环节中的常量折叠优化进行分析。

OptimizingCompiler 优化环节中的常量折叠优化具体指的是 kConstantFolding pass，它主要基于 SSA 形式中间代码进行常量折叠优化，它的具体实现类是 HconstantFolding。HconstantFolding 的定义和实现位于文件 art/compiler/optimizing/constant_floding.h 和

art/compiler/optimizing/constant_floding.cc 中。

kConstantFolding 在优化的过程中一共被调用了 4 次,在 5 个参数的 RunOptimizations() 函数(art/compiler/optimizing/optimizing_compiler.cc)中可以看出,这 4 次调用依据执行位置的不同将会优化不同的内容,这 4 次调用的代码如下:

```
//第 4 章/ optimizing_compiler.cc
void OptimizingCompiler::RunOptimizations(
    HGraph* graph, CodeGenerator* codegen,
    const DexCompilationUnit& dex_compilation_unit,
    PassObserver* pass_observer,
    VariableSizedHandleScope* handles) const {
  …
  OptimizationDef optimizations[] = {
    //Initial optimizations
    OptDef(OptimizationPass::kConstantFolding),
    OptDef(OptimizationPass::kInstructionSimplifier),
    OptDef(OptimizationPass::kDeadCodeElimination,
           "dead_code_elimination $ initial"),
    //Inlining
    OptDef(OptimizationPass::kInliner),
    //Simplification (only if inlining occurred)
    OptDef(OptimizationPass::kConstantFolding,
           "constant_folding $ after_inlining",
           OptimizationPass::kInliner),
    …
    //Simplification (TODO: only if GVN occurred)
    OptDef(OptimizationPass::kSelectGenerator),
    OptDef(OptimizationPass::kConstantFolding,
           "constant_folding $ after_gvn"),
    …
    //Simplification
    OptDef(OptimizationPass::kConstantFolding,
           "constant_folding $ after_bce"),
    …
```

在这几次调用中,除了第 1 次没有传递 pass 的 name 之外,其余 3 次都传递了不同的 pass name,这让相同的 pass 在不同的调用中可以进行区别。kConstantFolding 的第 1 次调用作为第 1 个运行的 pass 被调用;第 2 次调用依赖于 kInliner,并紧随其后被调用,所以 pass name 为 constant_folding $ after_inlining;第 3 次调用发生在 kGlobalValueNumbering 和 kSelectGenerator 之后,所以 pass name 为 constant_folding $ after_gvn;第 4 次调用发生在 kBoundsCheckElimination 和 kLoopOptimization 之后,所以 pass name 为 constant_folding $ after_bce。其实这几次不同的调用所采用的不同的名字,就是在 constant_folding $ 之后加上了位置相关信息,突出本次执行所在的位置。在不同的位置多次进行常量折叠优化,主要针对不同的位置所新产生出来的常量进行优化，并不是在同等条件下重复进行优化。

在 7 个参数的 RunOptimizations()函数（art/compiler/optimizing/optimizing_compiler.cc）中会对 pass 进行展开，执行 pass 所对应的具体实现类的对象。这个从 pass 到具体优化类的转换过程位于 art/compiler/optimizing/optimization.cc 中的 ConstructOptimizations()函数，因此，在整个的优化过程中，会有 4 个 HconstantFolding 的对象会被生成，它们对应不同的 pass name，分别为 constant_folding、constant_folding $ after_inlining、constant_folding $ after_gvn 和 constant_folding $ after_bce。

HconstantFolding 主要进行 SSA 形式 IR 的常量表达式优化，核心的算法位于 bool HConstantFolding::Run()函数内。Run()函数的实现位于 art/compiler/optimizing/constant_folding.cc 中，通过 HConstantFoldingVisitor 的 VisitReversePostOrder()函数完成工作，代码如下：

```
//第 4 章/ constant_folding.cc
bool HConstantFolding::Run() {
  HConstantFoldingVisitor visitor(graph_);
  //Process basic blocks in reverse post - order in the dominator tree
  //so that an instruction turned into a constant, used as input of
  //another instruction, may possibly be used to turn that second
  //instruction into a constant as well
  visitor.VisitReversePostOrder();
  return true;
}
```

Run()函数中的注释说得比较清楚，它对支配树（Dominator Tree）中的基本代码块（Basic Block）执行了逆后序（Reverse Post-order）处理，这样可以将一个常量计算指令转化为一个常量，然后将这个常量作为其他指令的输入，或许可以将其他指令也转化为常量。在这里需要特别注意，逆后序并不等于前序。逆后序需要在访问当前节点之前，访问当前节点的所有前驱节点。

Run()函数中所使用的 HConstantFoldingVisitor 的定义也位于 art/compiler/optimizing/constant_floding.cc 中，它是 HGraphDelegateVisitor 的子类，代码如下：

```
//第 4 章/ constant_folding.cc
//This visitor tries to simplify instructions that can be evaluated
//as constants
class HConstantFoldingVisitor : public HGraphDelegateVisitor {
public:
  explicit HConstantFoldingVisitor(HGraph * graph)
      : HGraphDelegateVisitor(graph) {}

private:
```

```
    void VisitBasicBlock(HBasicBlock * block) override;

    void VisitUnaryOperation(HUnaryOperation * inst) override;
    void VisitBinaryOperation(HBinaryOperation * inst) override;

    void VisitTypeConversion(HTypeConversion * inst) override;
    void VisitDivZeroCheck(HDivZeroCheck * inst) override;

    DISALLOW_COPY_AND_ASSIGN(HConstantFoldingVisitor);
};
```

其中,HGraphDelegateVisitor 是 HGraphVisitor 的子类,它的定义位于 art/compiler/optimizing/nodes.h 中,代码如下:

```
//第 4 章/ nodes.h
class HGraphDelegateVisitor : public HGraphVisitor {
public:
    explicit HGraphDelegateVisitor(HGraph * graph, OptimizingCompilerStats * stats = nullptr)
        : HGraphVisitor(graph, stats) {}
    virtual ~HGraphDelegateVisitor() {}

    //Visit functions that delegate to to super class
#define DECLARE_VISIT_INSTRUCTION(name, super)                            \
    void Visit##name(H##name * instr) override { Visit##super(instr); }

    FOR_EACH_INSTRUCTION(DECLARE_VISIT_INSTRUCTION)

#undef DECLARE_VISIT_INSTRUCTION

private:
    DISALLOW_COPY_AND_ASSIGN(HGraphDelegateVisitor);
};
```

继续向上追溯,HgraphVisitor 的定义也位于 nodes.h 中。HGraphDelegateVisitor 中会通过 FOR_EACH_INSTRUCTION(DECLARE_VISIT_INSTRUCTION)宏定义来创建一系列以 visit 开头的函数。

HconstantFolding 中的 Run()函数调用了 HConstantFoldingVisitor 的 VisitReversePostOrder()函数去实现所有操作,追溯 VisitReversePostOrder()函数的实现,可以从 HConstantFoldingVisitor 向上追溯到 HgraphVisitor 的 VisitReversePostOrder()函数,HConstantFoldingVisitor 及其父类 HGraphDelegateVisitor 都没有实现 VisitReversePostOrder()函数,所以 HConstantFoldingVisitor 所使用的 VisitReversePostOrder()函数是其父类 HGraphDelegateVisitor 的父类 HgraphVisitor 的 VisitReversePostOrder()函数,代码如下:

```
//第 4 章/ nodes.cc
void HGraphVisitor::VisitReversePostOrder() {
  for (HBasicBlock * block : graph_->GetReversePostOrder()) {
    VisitBasicBlock(block);
  }
}
```

这里根据逆后序遍历基本代码块，然后对每个基本代码块执行了 VisitBasicBlock() 函数。HConstantFoldingVisitor 类由自己的 VisitBasicBlock() 函数实现，位于 constant_floding.cc 中，代码如下：

```
//第 4 章/ constant_floding.cc
void HConstantFoldingVisitor::VisitBasicBlock(HBasicBlock * block) {
  //Traverse this block's instructions (phis don't need to be
  //processed) in (forward) order and replace the ones that can be
  //statically evaluated by a compile-time counterpart
  for (HInstructionIterator it(block->GetInstructions()); !it.Done();
       it.Advance()) {
    it.Current()->Accept(this);
  }
}
```

这里将遍历基本代码块中的每一条指令，然后执行其对应的 Accept() 函数。Accept() 函数在 HInstruction 类中被声明为虚函数，代码如下：

```
virtual void Accept(HGraphVisitor * visitor) = 0;
```

HInstruction 类的各个子类中也都有 Accept 的实现，是通过 art/compiler/optimizing/nodes.cc 中的宏实现的，代码如下：

```
//第 4 章/nodes.cc
#define DEFINE_ACCEPT(name, super)                    \
void H##name::Accept(HGraphVisitor * visitor)         \
{                                                     \
  visitor->Visit##name(this);                         \
}

FOR_EACH_CONCRETE_INSTRUCTION(DEFINE_ACCEPT)

#undef DEFINE_ACCEPT
```

这里每个 Accept 最终会调用 visitor 里最终的 Visit##name 形式的()函数，而 Visit##name 形式的()函数可以在 HGraphDelegateVisitor 类的定义中通过宏 FOR_EACH_

INSTRUCTION(DECLARE_VISIT_INSTRUCTION)被转化为 Visit##super(instr)的形式,其中 super 是 name 更上一层的描述,它们的对应关系可以在 art/compiler/optimizing/node.h 中的 FOR_EACH_CONCRETE_INSTRUCTION_COMMON 等宏内看到。除非 Visit##name 形式的()函数在 HConstantFoldingVisitor 类中已经有新的实现。

HConstantFoldingVisitor 类中,重写了 VisitUnaryOperation()函数和 VisitBinaryOperation()函数,它们属于 Visit##super 形式;重写了 VisitTypeConversion()函数和 VisitDivZeroCheck()函数,它们属于 Visit##name 形式。因为它们都是重写的,所以会覆盖父类的实现。正是这几个函数的操作完成了常量折叠优化。未重写的函数则会调用父类 HGraphDelegateVisitor 对应的 VisitXXXX()函数(这里包含了 Visit##super()形式的函数和 Visit##name()形式的函数),HGraphDelegateVisitor 中的 Visit##name()形式的函数调用最终会被转化为对应的 Visit##super()形式的函数调用,而最终 Visit##super()形式的函数会调用 HGraphVisitor 中的 VisitInstruction()函数,也就是执行了空操作,所以这个过程可以理解为,如果 HConstantFoldingVisitor 类中没有重写的 VisitXXX 函数,则最终都执行了一个空操作。

VisitBinaryOperation 中调用了 InstructionWithAbsorbingInputSimplifier()函数,它的实现也位于 constant_folding.cc 中,它是 HGraphVisitor 的子类,并且和 HConstantFoldingVisitor 在同一体系中,只不过它直接继承于 HGraphVisitor,而 HConstantFoldingVisitor 继承于 HGraphVisitor 的子类 HGraphDelegateVisitor。InstructionWithAbsorbingInputSimplifier 类重写了多个 Visit##name()形式的函数,这些 name 都属于 BinaryOperation 类别,所以最终这些函数也会被 VisitBinaryOperation()函数中的 inst—>Accept(&simplifier);所调用。这里同样涉及了 Accept。

本部分的优化算法并不复杂,复杂的是理清楚 3 个层次的 VisitXXX()函数的实现。第 1 个层次是 HGraphVisitor 层面的 Visit##name()函数和 Visit##super()函数,这一系列函数最终调用了 VisitInstruction()函数,实际上是个空操作;第 2 个层次是 HGraphVisitor 子类 HGraphDelegateVisitor 层面的 Visit##name()函数,它最终被转化成对应的 Visit##super()函数的调用;第 3 个层次是 HGraphDelegateVisitor 子类 HConstantFoldingVisitor 层面的 VisitUnaryOperation()函数、VisitBinaryOperation()函数、VisitTypeConversion()函数和 VisitDivZeroCheck()函数,它们属于对部分 Visit##super()函数和 Visit##name()函数的重写,前两者是 Visit##super()函数,后两者是 Visit##name()函数。等于只有第 3 个层次重写的函数是有效操作,其余的都是空操作,只是为了构建一个完整的体系。此外,InstructionWithAbsorbingInputSimplifier 作为 HGraphVisitor 的子类,也重写了一系列的 VisitXXX()函数,它们都属于 Visit##name()函数系列,并且这些 name 对应的 super 是 BinaryOperation。

需要指出的是,HGraphVisitor 和 HGraphDelegateVisitor 里的 VisitXXXX()系列函数的实现不太好理解,因为它们都是通过宏 FOR_EACH_INSTRUCTION(M)实现的大批量

定义。FOR_EACH_INSTRUCTION(M)又可以展开为 FOR_EACH_CONCRETE_INSTRUCTION(M)和 FOR_EACH_ABSTRACT_INSTRUCTION(M)。FOR_EACH_CONCRETE_INSTRUCTION(M)可以展开为与架构无关的 FOR_EACH_CONCRETE_INSTRUCTION_COMMON(M)、FOR_EACH_CONCRETE_INSTRUCTION_SHARED(M)和与架构相关的一系列宏。FOR_EACH_CONCRETE_INSTRUCTION_COMMON(M)展开后，所包含的(name，super)数据中的 super，正是 FOR_EACH_ABSTRACT_INSTRUCTION(M)所包含的(name，super)中的 name，所以根据宏生成的 VisitXXXX()函数之后，会覆盖所有的 name 和 super，除了最终的 Instruction，它是一个特例，在 HGraphVisitor 中由一个单独 VisitInstruction()函数实现，并不进行任何操作。在 FOR_EACH_CONCRETE_INSTRUCTION_SHARED(M)和与架构相关的一些宏中，其中的 name 所对应的 super，直接就是 Instruction。

FOR_EACH_INSTRUCTION(M)的相关实现位于 art/compiler/optimizing/nodes.h 中，代码如下：

```
//第 4 章/nodes.h
#define FOR_EACH_CONCRETE_INSTRUCTION_COMMON(M)              \
  M(Above, Condition)                                        \
  M(AboveOrEqual, Condition)                                 \
  M(Abs, UnaryOperation)                                     \
  M(Add, BinaryOperation)                                    \
  M(And, BinaryOperation)                                    \
  M(ArrayGet, Instruction)                                   \
  M(ArrayLength, Instruction)                                \
  M(ArraySet, Instruction)                                   \
  M(Below, Condition)                                        \
  M(BelowOrEqual, Condition)                                 \
  M(BooleanNot, UnaryOperation)                              \
  M(BoundsCheck, Instruction)                                \
  M(BoundType, Instruction)                                  \
  M(CheckCast, Instruction)                                  \
  M(ClassTableGet, Instruction)                              \
  M(ClearException, Instruction)                             \
  M(ClinitCheck, Instruction)                                \
  M(Compare, BinaryOperation)                                \
  M(ConstructorFence, Instruction)                           \
  M(CurrentMethod, Instruction)                              \
  M(ShouldDeoptimizeFlag, Instruction)                       \
  M(Deoptimize, Instruction)                                 \
  M(Div, BinaryOperation)                                    \
  M(DivZeroCheck, Instruction)                               \
  M(DoubleConstant, Constant)                                \
  M(Equal, Condition)                                        \
```

```
M(Exit, Instruction)                            \
M(FloatConstant, Constant)                      \
M(Goto, Instruction)                            \
M(GreaterThan, Condition)                       \
M(GreaterThanOrEqual, Condition)                \
M(If, Instruction)                              \
M(InstanceFieldGet, Instruction)                \
M(InstanceFieldSet, Instruction)                \
M(InstanceOf, Instruction)                      \
M(IntConstant, Constant)                        \
M(IntermediateAddress, Instruction)             \
M(InvokeUnresolved, Invoke)                     \
M(InvokeInterface, Invoke)                      \
M(InvokeStaticOrDirect, Invoke)                 \
M(InvokeVirtual, Invoke)                        \
M(InvokePolymorphic, Invoke)                    \
M(InvokeCustom, Invoke)                         \
M(LessThan, Condition)                          \
M(LessThanOrEqual, Condition)                   \
M(LoadClass, Instruction)                       \
M(LoadException, Instruction)                   \
M(LoadMethodHandle, Instruction)                \
M(LoadMethodType, Instruction)                  \
M(LoadString, Instruction)                      \
M(LongConstant, Constant)                       \
M(Max, Instruction)                             \
M(MemoryBarrier, Instruction)                   \
M(Min, BinaryOperation)                         \
M(MonitorOperation, Instruction)                \
M(Mul, BinaryOperation)                         \
M(NativeDebugInfo, Instruction)                 \
M(Neg, UnaryOperation)                          \
M(NewArray, Instruction)                        \
M(NewInstance, Instruction)                     \
M(Not, UnaryOperation)                          \
M(NotEqual, Condition)                          \
M(NullConstant, Instruction)                    \
M(NullCheck, Instruction)                       \
M(Or, BinaryOperation)                          \
M(PackedSwitch, Instruction)                    \
M(ParallelMove, Instruction)                    \
M(ParameterValue, Instruction)                  \
M(Phi, Instruction)                             \
M(Rem, BinaryOperation)                         \
M(Return, Instruction)                          \
M(ReturnVoid, Instruction)                      \
```

```
M(Ror, BinaryOperation)                                  \
M(Shl, BinaryOperation)                                  \
M(Shr, BinaryOperation)                                  \
M(StaticFieldGet, Instruction)                           \
M(StaticFieldSet, Instruction)                           \
M(UnresolvedInstanceFieldGet, Instruction)               \
M(UnresolvedInstanceFieldSet, Instruction)               \
M(UnresolvedStaticFieldGet, Instruction)                 \
M(UnresolvedStaticFieldSet, Instruction)                 \
M(Select, Instruction)                                   \
M(Sub, BinaryOperation)                                  \
M(SuspendCheck, Instruction)                             \
M(Throw, Instruction)                                    \
M(TryBoundary, Instruction)                              \
M(TypeConversion, Instruction)                           \
M(UShr, BinaryOperation)                                 \
M(Xor, BinaryOperation)                                  \
M(VecReplicateScalar, VecUnaryOperation)                 \
M(VecExtractScalar, VecUnaryOperation)                   \
M(VecReduce, VecUnaryOperation)                          \
M(VecCnv, VecUnaryOperation)                             \
M(VecNeg, VecUnaryOperation)                             \
M(VecAbs, VecUnaryOperation)                             \
M(VecNot, VecUnaryOperation)                             \
M(VecAdd, VecBinaryOperation)                            \
M(VecHalvingAdd, VecBinaryOperation)                     \
M(VecSub, VecBinaryOperation)                            \
M(VecMul, VecBinaryOperation)                            \
M(VecDiv, VecBinaryOperation)                            \
M(VecMin, VecBinaryOperation)                            \
M(VecMax, VecBinaryOperation)                            \
M(VecAnd, VecBinaryOperation)                            \
M(VecAndNot, VecBinaryOperation)                         \
M(VecOr, VecBinaryOperation)                             \
M(VecXor, VecBinaryOperation)                            \
M(VecSaturationAdd, VecBinaryOperation)                  \
M(VecSaturationSub, VecBinaryOperation)                  \
M(VecShl, VecBinaryOperation)                            \
M(VecShr, VecBinaryOperation)                            \
M(VecUShr, VecBinaryOperation)                           \
M(VecSetScalars, VecOperation)                           \
M(VecMultiplyAccumulate, VecOperation)                   \
M(VecSADAccumulate, VecOperation)                        \
M(VecDotProd, VecOperation)                              \
M(VecLoad, VecMemoryOperation)                           \
M(VecStore, VecMemoryOperation)                          \
```

```cpp
/*
 * Instructions, shared across several (not all) architectures.
 */
#if !defined(ART_ENABLE_CODEGEN_arm) && !defined(ART_ENABLE_CODEGEN_arm64)
#define FOR_EACH_CONCRETE_INSTRUCTION_SHARED(M)
#else
#define FOR_EACH_CONCRETE_INSTRUCTION_SHARED(M)                         \
  M(BitwiseNegatedRight, Instruction)                                   \
  M(DataProcWithShifterOp, Instruction)                                 \
  M(MultiplyAccumulate, Instruction)                                    \
  M(IntermediateAddressIndex, Instruction)
#endif

#define FOR_EACH_CONCRETE_INSTRUCTION_ARM(M)

#define FOR_EACH_CONCRETE_INSTRUCTION_ARM64(M)

#ifndef ART_ENABLE_CODEGEN_mips
#define FOR_EACH_CONCRETE_INSTRUCTION_MIPS(M)
#else
#define FOR_EACH_CONCRETE_INSTRUCTION_MIPS(M)                           \
  M(MipsComputeBaseMethodAddress, Instruction)                          \
  M(MipsPackedSwitch, Instruction)                                      \
  M(IntermediateArrayAddressIndex, Instruction)
#endif

#define FOR_EACH_CONCRETE_INSTRUCTION_MIPS64(M)

#ifndef ART_ENABLE_CODEGEN_x86
#define FOR_EACH_CONCRETE_INSTRUCTION_X86(M)
#else
#define FOR_EACH_CONCRETE_INSTRUCTION_X86(M)                            \
  M(X86ComputeBaseMethodAddress, Instruction)                           \
  M(X86LoadFromConstantTable, Instruction)                              \
  M(X86FPNeg, Instruction)                                              \
  M(X86PackedSwitch, Instruction)
#endif

#if defined(ART_ENABLE_CODEGEN_x86) || defined(ART_ENABLE_CODEGEN_x86_64)
#define FOR_EACH_CONCRETE_INSTRUCTION_X86_COMMON(M)                     \
  M(X86AndNot, Instruction)                                             \
  M(X86MaskOrResetLeastSetBit, Instruction)
#else
#define FOR_EACH_CONCRETE_INSTRUCTION_X86_COMMON(M)
#endif
```

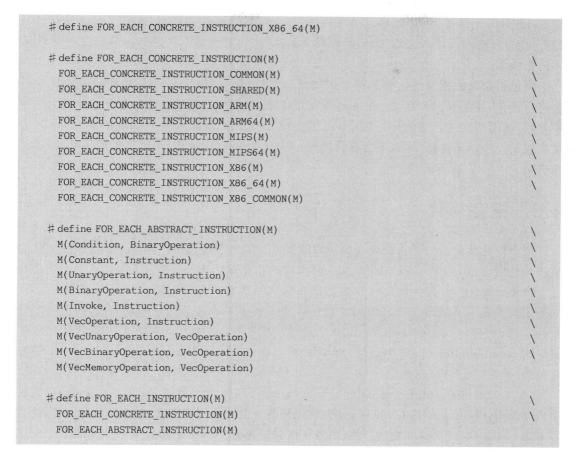

所以，HGraphVisitor 中其实具备了 Visit##name()函数、Visit##super()函数和 VisitInstruction()函数 3 个层面，其中 Visit##super()函数实际上是由 FOR_EACH_ABSTRACT_INSTRUCTION(M)宏中的 Visit##name()函数实现的，从 FOR_EACH_CONCRETE_INSTRUCTION_COMMON(M)宏的角度看，它就属于 Visit##super()函数了，也是为了不和 FOR_EACH_CONCRETE_INSTRUCTION_COMMON(M)的 Visit##name()函数重复，将其称为 Visit##super()函数，这里专门进行说明。也有一些指令的 super 直接是 Instruction，这种情况之下 Instruction 就是它的 super。在 HGraphVisitor 层面，所有的 VisitXXXX()函数最后都会调用 VisitInstruction()函数，然后不执行任何操作。到了 HGraphVisitor 的子类 HGraphDelegateVisitor 层面，它将对 Visit##name()函数的调用转化为对 Visit##super()函数的调用，将所有的调用层级上升了一个层面，所有指令都去访问其上一层，最终也都会执行 VisitInstruction()函数。

在 HGraphDelegateVisitor 子类 HConstantFoldingVisitor 层面上，重写了几个 VisitXXX()函数，这个层面保障的就是所有重写的函数都可以被执行，没有被重写的函数默认从父类继承（也就是空操作）。InstructionWithAbsorbingInputSimplifier 作为

HGraphVisitor 的子类,它重写了一系列的 VisitXXX() 函数,由于在这个层面也没有访问提升,所以所有重写的 VisitXXX() 函数都会被执行,没有被重写的函数就执行父类继承而来的空操作。

常量折叠是数据流优化的一种常用优化,并且可以在对 IR 进行不断优化的过程中多次进行优化,这主要因为在其他的一些优化过程中,会产生一些可以进行常量折叠优化的对象,所以需要多次进行该优化。这个优化的实现并不复杂,复杂在于 HGraphVisitor 及其子类构成的一个访问体系,这点不太容易理解,对于具体的优化动作,代码则较为简单,相对容易理解。

4.3 指令简化

指令简化优化指的是将指令在编译的时候进行简化替代,这样可以改进运行时的性能。本部分将对 OptimizingCompiler 的优化环节中的指令简化进行分析。

OptimizingCompiler 优化环节中的指令简化具体指的是 kInstructionSimplifier pass,它的实现类是 InstructionSimplifier。在针对 SSA 形式的图进行优化的过程中,如果优化对象为常量,则在 kConstantFolding pass 中进行;如果优化对象为非常数指令,则这种优化在 kInstructionSimplifier pass 中进行。本部分内容将对 kInstructionSimplifier pass 的实现进行分析。

从 art/compiler/optimizing/optimizing_compiler.cc 中 5 个参数的 RunOptimizations() 函数中可以看出,kInstructionSimplifier pass 在优化环节一共被执行了 5 次,其中前 4 次都是紧跟着 kConstantFolding pass 被执行的,最后 1 次是紧跟着 kCodeSinking pass 被执行的。

kInstructionSimplifier 紧跟着 kConstantFolding 执行很好理解,二者都是为了进行不同类型的指令优化,kInstructionSimplifier 用于处理非常数的情况,而 kConstantFolding 用于处理常数的情况,二者互补对所有指令都进行了优化。kCodeSinking pass 的情况在后面其他部分介绍。

InstructionSimplifier 的定义和实现位于 art/compiler/optimizing/instruction_simplifier.h 和 art/compiler/optimizing/ instruction_simplifier.cc 中。在 instruction_simplifier.cc 中,InstructionSimplifier 类有一个辅助它执行的 InstructionSimplifierVisitor 类,InstructionSimplifierVisitor 类也是 HGraphDelegateVisitor 的子类。

InstructionSimplifier 类的核心是 Run() 函数,它的 Run() 函数通过调用 InstructionSimplifierVisitor 类的 Run() 函数实现其优化,代码如下:

```
//第 4 章/instruction_simplifier.cc
bool InstructionSimplifier::Run() {
  if (kTestInstructionClonerExhaustively) {
    CloneAndReplaceInstructionVisitor visitor(graph_);
```

```
    visitor.VisitReversePostOrder();
  }

  InstructionSimplifierVisitor visitor(graph_, codegen_, stats_);
  return visitor.Run();
}
```

这里的核心代码是 InstructionSimplifierVisitor 的 Run() 函数,它通过逆后序访问基于 SSA 的图中的基本代码块,具体的代码如下:

```
//第 4 章/instruction_simplifier.cc
bool InstructionSimplifierVisitor::Run() {
  bool didSimplify = false;
  //Iterate in reverse post order to open up more simplifications to
  //users of instructions that got simplified
  for (HBasicBlock* block : GetGraph()->GetReversePostOrder()) {
    //The simplification of an instruction to another instruction may
    //yield possibilities for other simplifications. So although we
    //perform a reverse post order visit, we sometimes need to revisit
    //an instruction index
    do {
      simplification_occurred_ = false;
      VisitBasicBlock(block);
      if (simplification_occurred_) {
        didSimplify = true;
      }
    } while (simplification_occurred_ &&
             (simplifications_at_current_position_ < kMaxSamePositionSimplifications));
    simplifications_at_current_position_ = 0;
  }
  return didSimplify;
}
```

InstructionSimplifierVisitor 自身并没有实现 VisitBasicBlock() 函数,其父类 HGraphDelegateVisitor 也没有实现 VisitBasicBlock() 函数,但是 HGraphDelegateVisitor 的父类 HGraphVisitor 中有 VisitBasicBlock() 函数的实现,位于 art/compiler/optimizing/nodes.cc 中,主要通过 for 循环逐个访问基本代码块中的 phi 指令和其他指令,代码如下:

```
//第 4 章/nodes.cc
void HGraphVisitor::VisitBasicBlock(HBasicBlock* block) {
  for (HInstructionIterator it(block->GetPhis()); !it.Done();
       it.Advance()) {
    it.Current()->Accept(this);
  }
```

```
    for (HInstructionIterator it(block->GetInstructions()); !it.Done();
        it.Advance()) {
      it.Current()->Accept(this);
    }
}
```

InstructionSimplifierVisitor 的 Run() 函数实际上执行的是 HGraphVisitor 的 VisitBasicBlock() 函数，访问基本代码块中的 phi 指令和其他指令，并且执行每个指令的 Accept() 函数，Accept() 函数的实现情况在 HconstantFolding 中已经介绍过了，它会将操作转化为 InstructionSimplifierVisitor 类之内对访问每条指令的 VisitXXXX() 函数的调用。根据前文 HconstantFolding 中对 Visitor 体系的分析，这里也是类似的情况，只有 InstructionSimplifierVisitor 类重写的 VisitXXXX() 函数会被执行，其他指令的 VisitXXXX() 函数都会执行 HGraphVisitor 中的空操作。

InstructionSimplifierVisitor 类的 VisitXXXX() 函数一共有 39 个，代码如下：

```
//第 4 章/ instruction_simplifier.cc
    void VisitShift(HBinaryOperation* shift);
    void VisitEqual(HEqual* equal) override;
    void VisitNotEqual(HNotEqual* equal) override;
    void VisitBooleanNot(HBooleanNot* bool_not) override;
    void VisitInstanceFieldSet(HInstanceFieldSet* equal) override;
    void VisitStaticFieldSet(HStaticFieldSet* equal) override;
    void VisitArraySet(HArraySet* equal) override;
    void VisitTypeConversion(HTypeConversion* instruction) override;
    void VisitNullCheck(HNullCheck* instruction) override;
    void VisitArrayLength(HArrayLength* instruction) override;
    void VisitCheckCast(HCheckCast* instruction) override;
    void VisitAbs(HAbs* instruction) override;
    void VisitAdd(HAdd* instruction) override;
    void VisitAnd(HAnd* instruction) override;
    void VisitCondition(HCondition* instruction) override;
    void VisitGreaterThan(HGreaterThan* condition) override;
    void VisitGreaterThanOrEqual(HGreaterThanOrEqual* condition) override;
    void VisitLessThan(HLessThan* condition) override;
    void VisitLessThanOrEqual(HLessThanOrEqual* condition) override;
    void VisitBelow(HBelow* condition) override;
    void VisitBelowOrEqual(HBelowOrEqual* condition) override;
    void VisitAbove(HAbove* condition) override;
    void VisitAboveOrEqual(HAboveOrEqual* condition) override;
    void VisitDiv(HDiv* instruction) override;
    void VisitMul(HMul* instruction) override;
    void VisitNeg(HNeg* instruction) override;
    void VisitNot(HNot* instruction) override;
```

```
void VisitOr(HOr * instruction) override;
void VisitShl(HShl * instruction) override;
void VisitShr(HShr * instruction) override;
void VisitSub(HSub * instruction) override;
void VisitUShr(HUShr * instruction) override;
void VisitXor(HXor * instruction) override;
void VisitSelect(HSelect * select) override;
void VisitIf(HIf * instruction) override;
void VisitInstanceOf(HInstanceOf * instruction) override;
void VisitInvoke(HInvoke * invoke) override;
void VisitDeoptimize(HDeoptimize * deoptimize) override;
void VisitVecMul(HVecMul * instruction) override;
```

在这里的 VisitXXXX() 函数中，VisitCondition 和 VisitInvoke() 函数属于前文所提到的对 super 的访问，也就是说分别代表着对一类指令的访问，其他的 VisitXXXX() 函数都是对具体指令的访问，但是这里需要专门指出的是，对应着 super 为 Condition 的那些 name 所对应的 VisitXXXX() 函数在这里也专门进行了实现，一共有 10 个。其中，VisitEqual() 函数和 VisitNotEqual() 函数在进行了处理之后，再调用 VisitCondition() 函数；剩余的 8 个 name 所对应的 VisitXXXX() 函数则直接调用 VisitCondition() 函数，这 8 个函数的代码如下：

```
//第 4 章/ instruction_simplifier.cc
void InstructionSimplifierVisitor::VisitGreaterThan(
    HGreaterThan * condition) {
  VisitCondition(condition);
}

void InstructionSimplifierVisitor::VisitGreaterThanOrEqual(
    HGreaterThanOrEqual * condition) {
  VisitCondition(condition);
}

void InstructionSimplifierVisitor::VisitLessThan(HLessThan * condition) {
  VisitCondition(condition);
}

void InstructionSimplifierVisitor::VisitLessThanOrEqual(
    HLessThanOrEqual * condition) {
  VisitCondition(condition);
}

void InstructionSimplifierVisitor::VisitBelow(HBelow * condition) {
  VisitCondition(condition);
}
```

```
void InstructionSimplifierVisitor::VisitBelowOrEqual(
    HBelowOrEqual* condition) {
  VisitCondition(condition);
}

void InstructionSimplifierVisitor::VisitAbove(HAbove* condition) {
  VisitCondition(condition);
}

void InstructionSimplifierVisitor::VisitAboveOrEqual(
    HAboveOrEqual* condition) {
  VisitCondition(condition);
}
```

VisitCondition()函数也是 InstructionSimplifierVisitor 的成员函数,用于直接进行具体的优化操作,具体实现如下:

```
//第4章/ instruction_simplifier.cc
void InstructionSimplifierVisitor::VisitCondition(
    HCondition* condition) {
  if (condition->IsEqual() || condition->IsNotEqual()) {
    if (RecognizeAndSimplifyClassCheck(condition)) {
      return;
    }
  }

  //Reverse condition if left is constant. Our code generators prefer
  //constant on the right hand side
  if (condition->GetLeft()->IsConstant()
      && !condition->GetRight()->IsConstant()) {
    HBasicBlock* block = condition->GetBlock();
    HCondition* replacement =
        GetOppositeConditionSwapOps(block->GetGraph()->GetAllocator(),
                                    condition);
    //If it is a fp we must set the opposite bias
    if (replacement != nullptr) {
      if (condition->IsLtBias()) {
        replacement->SetBias(ComparisonBias::kGtBias);
      } else if (condition->IsGtBias()) {
        replacement->SetBias(ComparisonBias::kLtBias);
      }
      block->ReplaceAndRemoveInstructionWith(condition, replacement);
      RecordSimplification();

      condition = replacement;
```

```cpp
    }
}

HInstruction* left = condition->GetLeft();
HInstruction* right = condition->GetRight();

//Try to fold an HCompare into this HCondition

//We can only replace an HCondition which compares a Compare to 0
//Both 'dx' and 'jack' generate a compare to 0 when compiling a
//condition with a long, float or double comparison as input
if (!left->IsCompare() || !right->IsConstant() ||
    right->AsIntConstant()->GetValue() != 0) {
  //Conversion is not possible.
  return;
}

//Is the Compare only used for this purpose?
if (!left->GetUses().HasExactlyOneElement()) {
  //Someone else also wants the result of the compare.
  return;
}

if (!left->GetEnvUses().empty()) {
  //There is a reference to the compare result in an environment
  //Do we really need it?
  if (GetGraph()->IsDebuggable()) {
    return;
  }

  //We have to ensure that there are no deopt points in the sequence
  if (left->HasAnyEnvironmentUseBefore(condition)) {
    return;
  }
}

//Clean up any environment uses from the HCompare, if any
left->RemoveEnvironmentUsers();

//We have decided to fold the HCompare into the HCondition. Transfer
//the information
condition->SetBias(left->AsCompare()->GetBias());

//Replace the operands of the HCondition
condition->ReplaceInput(left->InputAt(0), 0);
condition->ReplaceInput(left->InputAt(1), 1);
```

```
    //Remove the HCompare
    left->GetBlock()->RemoveInstruction(left);

    RecordSimplification();
}
```

VisitInvoke()函数的实现,通过 switch-case 的结构,根据 Intrinsics 的类型进行各自不同的处理,代码如下:

```
//第 4 章/ instruction_simplifier.cc
void InstructionSimplifierVisitor::VisitInvoke(HInvoke* instruction) {
  switch (instruction->GetIntrinsic()) {
    case Intrinsics::kStringEquals:
      SimplifyStringEquals(instruction);
      break;
    case Intrinsics::kSystemArrayCopy:
      SimplifySystemArrayCopy(instruction);
      break;
    case Intrinsics::kIntegerRotateRight:
      SimplifyRotate(instruction, /* is_left = */ false,
                     DataType::Type::kInt32);
      break;
    case Intrinsics::kLongRotateRight:
      SimplifyRotate(instruction, /* is_left = */ false,
                     DataType::Type::kInt64);
      break;
    case Intrinsics::kIntegerRotateLeft:
      SimplifyRotate(instruction, /* is_left = */ true,
                     DataType::Type::kInt32);
      break;
    case Intrinsics::kLongRotateLeft:
      SimplifyRotate(instruction, /* is_left = */ true,
                     DataType::Type::kInt64);
      break;
    case Intrinsics::kIntegerCompare:
      SimplifyCompare(instruction, /* is_signum = */ false,
                      DataType::Type::kInt32);
      break;
    case Intrinsics::kLongCompare:
      SimplifyCompare(instruction, /* is_signum = */ false,
                      DataType::Type::kInt64);
      break;
    case Intrinsics::kIntegerSignum:
      SimplifyCompare(instruction, /* is_signum = */ true,
                      DataType::Type::kInt32);
```

```cpp
    break;
  case Intrinsics::kLongSignum:
    SimplifyCompare(instruction, /* is_signum = */ true,
                    DataType::Type::kInt64);
    break;
  case Intrinsics::kFloatIsNaN:
  case Intrinsics::kDoubleIsNaN:
    SimplifyIsNaN(instruction);
    break;
  case Intrinsics::kFloatFloatToIntBits:
  case Intrinsics::kDoubleDoubleToLongBits:
    SimplifyFP2Int(instruction);
    break;
  case Intrinsics::kStringCharAt:
    SimplifyStringCharAt(instruction);
    break;
  case Intrinsics::kStringIsEmpty:
  case Intrinsics::kStringLength:
    SimplifyStringIsEmptyOrLength(instruction);
    break;
  case Intrinsics::kStringIndexOf:
  case Intrinsics::kStringIndexOfAfter:
    SimplifyStringIndexOf(instruction);
    break;
  case Intrinsics::kStringStringIndexOf:
  case Intrinsics::kStringStringIndexOfAfter:
    SimplifyNPEOnArgN(instruction, 1); //0th has own NullCheck
    break;
  case Intrinsics::kStringBufferAppend:
  case Intrinsics::kStringBuilderAppend:
    SimplifyReturnThis(instruction);
    break;
  case Intrinsics::kStringBufferToString:
  case Intrinsics::kStringBuilderToString:
    SimplifyAllocationIntrinsic(instruction);
    break;
  case Intrinsics::kUnsafeLoadFence:
    SimplifyMemBarrier(instruction, MemBarrierKind::kLoadAny);
    break;
  case Intrinsics::kUnsafeStoreFence:
    SimplifyMemBarrier(instruction, MemBarrierKind::kAnyStore);
    break;
  case Intrinsics::kUnsafeFullFence:
    SimplifyMemBarrier(instruction, MemBarrierKind::kAnyAny);
    break;
  case Intrinsics::kVarHandleFullFence:
```

```cpp
    SimplifyMemBarrier(instruction, MemBarrierKind::kAnyAny);
    break;
  case Intrinsics::kVarHandleAcquireFence:
    SimplifyMemBarrier(instruction, MemBarrierKind::kLoadAny);
    break;
  case Intrinsics::kVarHandleReleaseFence:
    SimplifyMemBarrier(instruction, MemBarrierKind::kAnyStore);
    break;
  case Intrinsics::kVarHandleLoadLoadFence:
    SimplifyMemBarrier(instruction, MemBarrierKind::kLoadAny);
    break;
  case Intrinsics::kVarHandleStoreStoreFence:
    SimplifyMemBarrier(instruction, MemBarrierKind::kStoreStore);
    break;
  case Intrinsics::kMathMinIntInt:
    SimplifyMin(instruction, DataType::Type::kInt32);
    break;
  case Intrinsics::kMathMinLongLong:
    SimplifyMin(instruction, DataType::Type::kInt64);
    break;
  case Intrinsics::kMathMinFloatFloat:
    SimplifyMin(instruction, DataType::Type::kFloat32);
    break;
  case Intrinsics::kMathMinDoubleDouble:
    SimplifyMin(instruction, DataType::Type::kFloat64);
    break;
  case Intrinsics::kMathMaxIntInt:
    SimplifyMax(instruction, DataType::Type::kInt32);
    break;
  case Intrinsics::kMathMaxLongLong:
    SimplifyMax(instruction, DataType::Type::kInt64);
    break;
  case Intrinsics::kMathMaxFloatFloat:
    SimplifyMax(instruction, DataType::Type::kFloat32);
    break;
  case Intrinsics::kMathMaxDoubleDouble:
    SimplifyMax(instruction, DataType::Type::kFloat64);
    break;
  case Intrinsics::kMathAbsInt:
    SimplifyAbs(instruction, DataType::Type::kInt32);
    break;
  case Intrinsics::kMathAbsLong:
    SimplifyAbs(instruction, DataType::Type::kInt64);
    break;
  case Intrinsics::kMathAbsFloat:
    SimplifyAbs(instruction, DataType::Type::kFloat32);
```

```
      break;
    case Intrinsics::kMathAbsDouble:
      SimplifyAbs(instruction, DataType::Type::kFloat64);
      break;
    default:
      break;
  }
}
```

这里 Intrinsic 的枚举类型的定义位于 art/Runtime/intrinsics_enum.h 中,代码如下:

```
//第 4 章/ intrinsics_enum.h
enum class Intrinsics {
#define OPTIMIZING_INTRINSICS(Name, ...) \
  k ## Name,
#include "intrinsics_list.h"
  kNone,
  INTRINSICS_LIST(OPTIMIZING_INTRINSICS)
#undef INTRINSICS_LIST
#undef OPTIMIZING_INTRINSICS
};
std::ostream& operator <<(std::ostream& os, const Intrinsics& intrinsic);

} //namespace art
```

其中所涉及的 INTRINSICS_LIST 宏的定位位于 art/Runtime/intrinsics_list.h 中。

至此,InstructionSimplifier 及其相关的 InstructionSimplifierVisitor 的结构就比较清晰了,其余的分析都是针对具体的指令在操作过程中所进行的具体优化,可以根据指令名称直接分析 Visit##name()函数的内容。由于 InstructionSimplifier 和 HconstantFolding 的整体结构类似,并且二者在实际的执行过程中是相辅相成的,所以将两者放到一起讲解更加利于理解。

本部分对于指令简化的具体优化进行了分析,它和之前介绍的常量折叠相似,二者在处理的结构上也相似,可以将二者对照进行理解。

4.4 死代码优化

死代码优化通常是指优化没有实际意义的代码。OptimizingCompiler 优化环节中的死代码优化所对应的是 kDeadCodeElimination pass,它的主要作用是基于 SSA 去消除不使用的死变量或者死指令,它的实现类是 HDeadCodeElimination。本部分将对 HDeadCodeElimination 及其相关的实现进行分析。

根据 art/compiler/optimizing/optimizing_compiler.cc 中 5 个参数的 RunOptimizations()函

数中的 pass 序列，kDeadCodeElimination pass 在优化的过程中一共被执行了 4 次，前 3 次都是紧跟着 kConstantFolding pass 和 kInstructionSimplifier 被执行的，只有最后 1 次不同，是跟随着 kCHAGuardOptimization 被执行的，代码如下：

```cpp
//第 4 章/ optimizing_compiler.cc
void OptimizingCompiler::RunOptimizations(
    HGraph* graph, CodeGenerator* codegen,
    const DexCompilationUnit& dex_compilation_unit,
    PassObserver* pass_observer,
    VariableSizedHandleScope* handles) const {
…

  OptimizationDef optimizations[] = {
    //Initial optimizations
    OptDef(OptimizationPass::kConstantFolding),
    OptDef(OptimizationPass::kInstructionSimplifier),
    OptDef(OptimizationPass::kDeadCodeElimination,
           "dead_code_elimination $ initial"),
    //Inlining
    OptDef(OptimizationPass::kInliner),
    //Simplification (only if inlining occurred)
    OptDef(OptimizationPass::kConstantFolding,
           "constant_folding $ after_inlining",
           OptimizationPass::kInliner),
    OptDef(OptimizationPass::kInstructionSimplifier,
           "instruction_simplifier $ after_inlining",
           OptimizationPass::kInliner),
    OptDef(OptimizationPass::kDeadCodeElimination,
           "dead_code_elimination $ after_inlining",
           OptimizationPass::kInliner),
    //GVN
    OptDef(OptimizationPass::kSideEffectsAnalysis,
           "side_effects $ before_gvn"),
    OptDef(OptimizationPass::kGlobalValueNumbering),
    //Simplification (TODO: only if GVN occurred)
    OptDef(OptimizationPass::kSelectGenerator),
    OptDef(OptimizationPass::kConstantFolding,
           "constant_folding $ after_gvn"),
    OptDef(OptimizationPass::kInstructionSimplifier,
           "instruction_simplifier $ after_gvn"),
    OptDef(OptimizationPass::kDeadCodeElimination,
           "dead_code_elimination $ after_gvn"),
    //High-level optimizations
    OptDef(OptimizationPass::kSideEffectsAnalysis,
           "side_effects $ before_licm"),
```

```
        OptDef(OptimizationPass::kInvariantCodeMotion),
        OptDef(OptimizationPass::kInductionVarAnalysis),
        OptDef(OptimizationPass::kBoundsCheckElimination),
        OptDef(OptimizationPass::kLoopOptimization),
        //Simplification
        OptDef(OptimizationPass::kConstantFolding,
                "constant_folding$after_bce"),
        OptDef(OptimizationPass::kInstructionSimplifier,
                "instruction_simplifier$after_bce"),
        //Other high-level optimizations
        OptDef(OptimizationPass::kSideEffectsAnalysis,
                "side_effects$before_lse"),
        OptDef(OptimizationPass::kLoadStoreAnalysis),
        OptDef(OptimizationPass::kLoadStoreElimination),
        OptDef(OptimizationPass::kCHAGuardOptimization),
        OptDef(OptimizationPass::kDeadCodeElimination,
                "dead_code_elimination$final"),
        ...
    }
```

HDeadCodeElimination 作为 kDeadCodeElimination pass 的实现类，它的定义和实现位于 art/compiler/optimizing/dead_code_elimination.h 和 art/compiler/optimizing/dead_code_elimination.cc 中。与之前 pass 的实现类的情况不同，HDeadCodeElimination 并没有配套的 XXXXVistor 类来辅助执行，核心代码还是在 HDeadCodeElimination 的 Run()函数内，代码如下：

```
//第 4 章/dead_code_elimination.cc
bool HDeadCodeElimination::Run() {
    //Do not eliminate dead blocks if the graph has irreducible loops. We
    //could support it, but that would require changes in our loop
    //representation to handle multiple entry points. We decided it was
    //not worth the complexity
    if (!graph_->HasIrreducibleLoops()) {
        //Simplify graph to generate more dead block patterns
        ConnectSuccessiveBlocks();
        bool did_any_simplification = false;
        did_any_simplification |= SimplifyAlwaysThrows();
        did_any_simplification |= SimplifyIfs();
        did_any_simplification |= RemoveDeadBlocks();
        if (did_any_simplification) {
            //Connect successive blocks created by dead branches
            ConnectSuccessiveBlocks();
        }
    }
```

```
    SsaRedundantPhiElimination(graph_).Run();
    RemoveDeadInstructions();
    return true;
}
```

Run()函数中并没有直接处理 SSA 图中有不可简化的循环情况下的死代码块,觉得这部分内容会增加复杂度,有点得不偿失,所以 Run()函数的内部被分为两部分,一部分是不具有不可简化循环的处理,这部分内容是在 if(! graph_->HasIrreducibleLoops())这个判断条件内执行,这部分内容执行了 SimplifyAlwaysThrows()函数、SimplifyIfs()函数和 RemoveDeadBlocks()函数;另一部分是直接调用了 SsaRedundantPhiElimination 类的 Run()函数,然后执行了 RemoveDeadInstructions()函数。这里所提到的 SsaRedundantPhiElimination 类的定义和实现位于 art/compiler/optimizing/ssa_phi_elimination.h 和 art/compiler/optimizing/ssa_phi_elimination.cc 中,它也是 HOptimization 的子类,主要用来移除多余的 phi 节点,这些多余的 phi 节点是在 SSA 转换的过程中添加的。

在 Run()函数中,还执行了两次 ConnectSuccessiveBlocks()函数,这个函数在未优化之前被执行,这样可以简化树的结构,方便更好地发现死代码块;在优化之后调用这个函数,可以将一些死分支所造成的代码块连接到一起。SimplifyAlwaysThrows()函数、SimplifyIfs()函数和 RemoveDeadBlocks()函数,每个函数都专门用于处理一种情况。SimplifyAlwaysThrows()函数主要对总是抛出异常情况下图的模式进行转换,这点在代码的注释中有解释,具体的代码如下:

```
//第 4 章/dead_code_elimination.cc
//Simplify the pattern:
//
//     B1
//    / \
//   |  foo()   //always throws
//   \  goto B2
//    \ /
//     B2
//
//Into:
//
//     B1
//    / \
//   |  foo()
//   |  goto Exit
//   |   |
//   B2 Exit
//
//Rationale
```

```
//Removal of the never taken edge to B2 may expose
//other optimization opportunities, such as code sinking
```

SimplifyIfs()函数可以简化 if 所构成的模块分割模式,所进行的模式转化在代码注释中也有介绍,具体如下:

```
//第 4 章/dead_code_elimination.cc
//Simplify the pattern:
//
// B1    B2   ...
// goto goto goto
//   \   |   /
//    \  |  /
//       B3
//    i1 = phi(input, input)
//    (i2 = condition on i1)
//    if i1 (or i2)
//      /   \
//     /     \
//    B4     B5
//
//Into:
//
// B1    B2   ...
// |     |    |
// B4    B5   B?
//
//Note that individual edges can be redirected (for example B2 -> B3
//can be redirected as B2 -> B5) without applying this optimization
//to other incoming edges
```

同时,这个函数所进行的简化工作并不适用于 catch 基本代码块。

RemoveDeadBlocks()函数则用于遍历所有的代码块,将其中间不可达的代码块删除,并且在删除代码块之后,还要重新进行整个支配树的计算。

SsaRedundantPhiElimination 类的 Run()函数的调用,是通过传递给 SsaRedundant-PhiElimination 类 graph_ 创建出一个对象,然后调用这个对象的 Run()函数。这个 Run()函数的具体实现位于 art/compiler/optimizing/ssa_phi_elimination.cc 中,主要用于清理多余的 phi 节点。

RemoveDeadInstructions()函数则通过遍历基本代码块中所有的指令,对每条指令都做是不是死指令且可以移除的判断,最终将可以移除的指令都移除。从这部分代码比较容易看出如何清理死指令,适合初学者理解。具体代码如下:

```cpp
//第 4 章/dead_code_elimination.cc
void HDeadCodeElimination::RemoveDeadInstructions() {
  //Process basic blocks in post-order in the dominator tree, so that
  //a dead instruction depending on another dead instruction is removed
  for (HBasicBlock* block : graph_->GetPostOrder()) {
    //Traverse this block's instructions in backward order and remove
    //the unused ones
    HBackwardInstructionIterator i(block->GetInstructions());
    //Skip the first iteration, as the last instruction of a block is
    //a branching instruction
    DCHECK(i.Current()->IsControlFlow());
    for (i.Advance(); !i.Done(); i.Advance()) {
      HInstruction* inst = i.Current();
      DCHECK(!inst->IsControlFlow());
      if (inst->IsDeadAndRemovable()) {
        block->RemoveInstruction(inst);
        MaybeRecordStat(stats_,
            MethodCompilationStat::kRemovedDeadInstruction);
      }
    }
  }
}
```

HDeadCodeElimination 类和之前的 HconstantFolding 和 InstructionSimplifier 不太相同，它没有配套的 XXXXVisitor 去帮助它执行 Run() 函数的核心内容，而是通过自己在 Run() 函数之内执行了主要的操作，并且还调用了 HOptimization 的另外一个子类 SsaRedundantPhiElimination 的 Run() 函数。

本部分介绍了 OptimizingCompiler 优化环节中的死代码优化，死代码优化是编译器优化中比较常见的优化，也比较容易理解，可以作为初学者深入理解优化的入口内容。

4.5 循环体优化

循环体优化指的是嵌套循环的死指令和空循环移除、内部循环向量化等优化动作，旨在提高执行效率。本部分将对 OptimizingCompiler 优化环节中的循环体优化进行分析。

OptimizingCompiler 优化环节中的循环体优化具体指的是 kLoopOptimization pass，它在整体的优化过程中只被执行了一次，它所对应的具体实现类是 HloopOptimization。HloopOptimization 是 HOptimization 的子类，它的实现位于 art/compiler/optimizing/loop_optimization.h 和 art/compiler/optimizing/loop_optimization.cc 中。HloopOptimization 的优化操作也是在 Run() 函数中执行的，代码如下：

```
//第4章/loop_optimization.cc
bool HLoopOptimization::Run() {
  //Skip if there is no loop or the graph has try-catch/irreducible
  //loops. TODO: make this less of a sledgehammer
  if (!graph_->HasLoops() || graph_->HasTryCatch() ||
      graph_->HasIrreducibleLoops()) {
    return false;
  }

  //Phase-local allocator
  ScopedArenaAllocator allocator(graph_->GetArenaStack());
  loop_allocator_ = &allocator;

  //Perform loop optimizations
  bool didLoopOpt = LocalRun();
  if (top_loop_ == nullptr) {
    graph_->SetHasLoops(false); //no more loops
  }

  //Detach
  loop_allocator_ = nullptr;
  last_loop_ = top_loop_ = nullptr;

  return didLoopOpt;
}
```

其中,核心的函数是 LocalRun() 函数,通过它进行循环优化,LocalRun() 函数的实现也位于 loop_optimization.cc 中,代码如下:

```
//第4章/loop_optimization.cc
//
//Loop setup and traversal
//

bool HLoopOptimization::LocalRun() {
  bool didLoopOpt = false;
  //Build the linear order using the phase-local allocator. This step
  //enables building a loop hierarchy that properly reflects the
  //outer-inner and previous-next relation
  ScopedArenaVector<HBasicBlock*> linear_order(
                         loop_allocator_->Adapter(kArenaAllocLinearOrder));
  LinearizeGraph(graph_, &linear_order);

  //Build the loop hierarchy
  for (HBasicBlock* block : linear_order) {
```

```
    if (block->IsLoopHeader()) {
      AddLoop(block->GetLoopInformation());
    }
  }

  //Traverse the loop hierarchy inner-to-outer and optimize. Traversal
  //can use temporary data structures using the phase-local allocator
  //All new HIR should use the global allocator
  if (top_loop_ != nullptr) {
    ScopedArenaSet<HInstruction*> iset(
                    loop_allocator_->Adapter(kArenaAllocLoopOptimization));
    ScopedArenaSafeMap<HInstruction*, HInstruction*> reds(
                                                std::less<HInstruction*>(),
                    loop_allocator_->Adapter(kArenaAllocLoopOptimization));
    ScopedArenaSet<ArrayReference> refs(
                    loop_allocator_->Adapter(kArenaAllocLoopOptimization));
    ScopedArenaSafeMap<HInstruction*, HInstruction*> map(
                                                std::less<HInstruction*>(),
                    loop_allocator_->Adapter(kArenaAllocLoopOptimization));
    ScopedArenaSafeMap<HInstruction*, HInstruction*> perm(
                                                std::less<HInstruction*>(),
                    loop_allocator_->Adapter(kArenaAllocLoopOptimization));
    //Attach
    iset_ = &iset;
    reductions_ = &reds;
    vector_refs_ = &refs;
    vector_map_ = &map;
    vector_permanent_map_ = &perm;
    //Traverse
    didLoopOpt = TraverseLoopsInnerToOuter(top_loop_);
    //Detach
    iset_ = nullptr;
    reductions_ = nullptr;
    vector_refs_ = nullptr;
    vector_map_ = nullptr;
    vector_permanent_map_ = nullptr;
  }
  return didLoopOpt;
}
```

在这里首先将图线性化，然后构建出 loop 的结构，确定其中的 loop 信息，之后通过 TraverseLoopsInnerToOuter() 函数从内到外遍历 loop，进行优化，所以 TraverseLoopsInnerToOuter() 函数是 LocalRun() 函数的核心操作，具体的优化环节都在这里进行。TraverseLoopsInnerToOuter() 函数的实现位于 loop_optimization.cc 中，代码如下：

```cpp
//第 4 章/loop_optimization.cc
bool HLoopOptimization::TraverseLoopsInnerToOuter(LoopNode * node) {
  bool changed = false;
  for ( ; node != nullptr; node = node->next) {
    //Visit inner loops first. Recompute induction information for this
    //loop if the induction of any inner loop has changed
    if (TraverseLoopsInnerToOuter(node->inner)) {
      induction_range_.ReVisit(node->loop_info);
      changed = true;
    }
    //Repeat simplifications in the loop-body until no more changes
    //occur. Note that since each simplification consists of eliminating
    //code (without introducing new code), this process is always
    //finite
    do {
      simplified_ = false;
      SimplifyInduction(node);
      SimplifyBlocks(node);
      changed = simplified_ || changed;
    } while (simplified_);
    //Optimize inner loop
    if (node->inner == nullptr) {
      changed = OptimizeInnerLoop(node) || changed;
    }
  }
  return changed;
}
```

通过遍历 LoopNode 指针类型的 node，执行 SimplifyInduction()、SimplifyBlocks() 及 OptimizeInnerLoop() 这 3 个函数。SimplifyInduction() 函数负责简化 induction，SimplifyBlocks() 函数负责简化基本代码块，OptimizeInnerLoop() 函数用于对最内层的 loop 进行优化。

SimplifyInduction()函数主要负责简化 Induction Cycle，这些 Induction Cycle 涉及的变量不会在循环内部使用。主要方式是通过遍历 loop 的 header 节点中的 phi 信息，根据 phi 的 induction 信息找出在循环内部并不使用的变量，然后移除 Induction Cycle，并且将变量的最新值更新到循环后使用的地方。有的时候，也会出现循环内部的 Induction Cycle 并没有去除，但是变量的最新值也更新到了循环后使用的地方，这两者并不会冲突，代码如下：

```cpp
//第 4 章/loop_optimization.cc
void HLoopOptimization::SimplifyInduction(LoopNode * node) {
  HBasicBlock * header = node->loop_info->GetHeader();
  HBasicBlock * preheader = node->loop_info->GetPreHeader();
  //Scan the phis in the header to find opportunities to simplify an
```

```
//induction cycle that is only used outside the loop. Replace these
//uses, if any, with the last value and remove the induction cycle
//Examples: for (int i = 0; x != null; i++) { .... no i .... }
// for (int i = 0; i < 10; i++, k++) { .... no k .... } return k;
for (HInstructionIterator it(header->GetPhis()); !it.Done();
     it.Advance()) {
  HPhi* phi = it.Current()->AsPhi();
  if (TrySetPhiInduction(phi, /*restrict_uses*/ true) &&
      TryAssignLastValue(node->loop_info, phi, preheader,
      /*collect_loop_uses*/ false)) {
    //Note that it's ok to have replaced uses after the loop with the
    //last value, without being able to remove the cycle. Environment
    //uses (which are the reason we may not be able to remove the
    //cycle) within the loop will still hold the right value. We must
    //have tried first, however, to replace outside uses
    if (CanRemoveCycle()) {
      simplified_ = true;
      for (HInstruction* i : *iset_) {
        RemoveFromCycle(i);
      }
      DCHECK(CheckInductionSetFullyRemoved(iset_));
    }
  }
}
```

SimplifyBlocks()函数主要通过遍历所有的循环体的基本代码块,移除循环体中的死指令和琐碎的控制流基本代码块,代码如下:

```
//第 4 章/loop_optimization.cc
void HLoopOptimization::SimplifyBlocks(LoopNode* node) {
  //Iterate over all basic blocks in the loop-body
  for (HBlocksInLoopIterator it(*node->loop_info); !it.Done();
       it.Advance()) {
    HBasicBlock* block = it.Current();
    //Remove dead instructions from the loop-body
    RemoveDeadInstructions(block->GetPhis());
    RemoveDeadInstructions(block->GetInstructions());
    //Remove trivial control flow blocks from the loop-body
    if (block->GetPredecessors().size() == 1 &&
        block->GetSuccessors().size() == 1 &&
        block->GetSingleSuccessor()->GetPredecessors().size() == 1) {
      simplified_ = true;
      block->MergeWith(block->GetSingleSuccessor());
    } else if (block->GetSuccessors().size() == 2) {
```

```
        //Trivial if block can be bypassed to either branch
        HBasicBlock* succ0 = block->GetSuccessors()[0];
        HBasicBlock* succ1 = block->GetSuccessors()[1];
        HBasicBlock* meet0 = nullptr;
        HBasicBlock* meet1 = nullptr;
        if (succ0 != succ1 &&
            IsGotoBlock(succ0, &meet0) &&
            IsGotoBlock(succ1, &meet1) &&
            meet0 == meet1 &&                    //meets again
            meet0 != block &&                    //no self-loop
            meet0->GetPhis().IsEmpty()) {        //not used for merging
          simplified_ = true;
          succ0->DisconnectAndDelete();
          if (block->Dominates(meet0)) {
            block->RemoveDominatedBlock(meet0);
            succ1->AddDominatedBlock(meet0);
            meet0->SetDominator(succ1);
          }
        }
      }
    }
  }
```

OptimizeInnerLoop()函数则通过TryOptimizeInnerLoopFinite()函数和TryPeelingAndUnrolling()函数实现其功能，代码如下：

```
//第4章/loop_optimization.cc
bool HLoopOptimization::OptimizeInnerLoop(LoopNode* node) {
  return TryOptimizeInnerLoopFinite(node) ||
         TryPeelingAndUnrolling(node);
}
```

TryOptimizeInnerLoopFinite()函数对符合条件的内部循环进行优化，可以移除空循环或者只循环一次琐碎的循环，并且在可能的情况下对循环进行向量化。TryPeelingAndUnrolling()函数则对循环进行Peeling和Unrolling。

SimplifyInduction()、SimplifyBlocks()和OptimizeInnerLoop()这3个函数，是HloopOptimization类中的重要成员函数，分别代表循环优化的一大部分，对于其中的具体细节，可以从函数中进行更加深入的跟踪，在此不再介绍。

4.6 指令下沉

指令下沉优化具体指的是kCodeSinking pass，它主要负责将指令尽可能地移动到非公共分支，减小公共分支指令执行不必要的开销，它所对应的实现类为CodeSinking。本部分将对 CodeSinking 的实现进行分析。

CodeSinking 的定义和实现位于 art/compiler/optimizing/code_sinking.h 和 art/compiler/optimizing/code_sinking.cc 中。CodeSinking 的优化相关操作位于 Run() 函数中,Run() 函数位于 code_sinking.cc 中,代码如下:

```
//第 4 章/code_sinking.cc
bool CodeSinking::Run() {
  HBasicBlock* exit = graph_->GetExitBlock();
  if (exit == nullptr) {
    //Infinite loop, just bail
    return false;
  }
  //TODO(ngeoffray): we do not profile branches yet, so use throw
  //instructions as an indicator of an uncommon branch
  for (HBasicBlock* exit_predecessor : exit->GetPredecessors()) {
    HInstruction* last = exit_predecessor->GetLastInstruction();
    //Any predecessor of the exit that does not return, throws an
    //exception
    if (!last->IsReturn() && !last->IsReturnVoid()) {
      SinkCodeToUncommonBranch(exit_predecessor);
    }
  }
  return true;
}
```

Run() 函数通过 for 循环,遍历访问所有 exit 基本代码块的所有前驱基本代码块,然后在基本代码块的最后一条指令不是 return 且不是 reture void 的情况下,对该基本代码块执行 SinkCodeToUncommonBranch() 函数。SinkCodeToUncommonBranch() 函数是 Run() 函数的核心,也是 CodeSinking 仅有的两个成员函数之一,它的功能是把只被最后基本代码块、post-dominated 基本代码块和 dominated 基本代码块所使用的代码移动到这些基本代码块中,代码如下:

```
//第 4 章/code_sinking.cc
void CodeSinking::SinkCodeToUncommonBranch(HBasicBlock* end_block) {
  //Local allocator to discard data structures created below at the end
  //of this optimization
  ScopedArenaAllocator allocator(graph_->GetArenaStack());

  size_t number_of_instructions = graph_->GetCurrentInstructionId();
  ScopedArenaVector<HInstruction*>
                                worklist(allocator.Adapter(kArenaAllocMisc));
  ArenaBitVector processed_instructions(&allocator,
```

```cpp
                                        number_of_instructions,
                                        /* expandable= */ false);
processed_instructions.ClearAllBits();
ArenaBitVector post_dominated(&allocator, graph_->GetBlocks().size(),
                              /* expandable= */ false);
post_dominated.ClearAllBits();
ArenaBitVector instructions_that_can_move(&allocator,
                                          number_of_instructions,
                                          /* expandable= */ false);
instructions_that_can_move.ClearAllBits();
ScopedArenaVector<HInstruction*>
                    move_in_order(allocator.Adapter(kArenaAllocMisc));

//Step (1): Visit post order to get a subset of blocks post dominated
//by `end_block`
//TODO(ngeoffray): Getting the full set of post-dominated shoud be
//done by computint the post dominator tree, but that could be too
//time consuming. Also, we should start the analysis from blocks
//dominated by an uncommon branch, but we don't profile branches yet
bool found_block = false;
for (HBasicBlock* block : graph_->GetPostOrder()) {
  if (block == end_block) {
    found_block = true;
    post_dominated.SetBit(block->GetBlockId());
  } else if (found_block) {
    bool is_post_dominated = true;
    if (block->GetSuccessors().empty()) {
      //We currently bail for loops
      is_post_dominated = false;
    } else {
      for (HBasicBlock* successor : block->GetSuccessors()) {
        if (!post_dominated.IsBitSet(successor->GetBlockId())) {
          is_post_dominated = false;
          break;
        }
      }
    }
    if (is_post_dominated) {
      post_dominated.SetBit(block->GetBlockId());
    }
  }
}

//Now that we have found a subset of post-dominated blocks, add to the
//worklist all inputs of instructions in these blocks that are not
//themselves in these blocks
```

```cpp
//Also find the common dominator of the found post dominated blocks
//to help filtering out un-movable uses in step (2)
CommonDominator finder(end_block);
for (size_t i = 0, e = graph_->GetBlocks().size(); i < e; ++i) {
  if (post_dominated.IsBitSet(i)) {
    finder.Update(graph_->GetBlocks()[i]);
    AddInputs(graph_->GetBlocks()[i], processed_instructions,
              post_dominated, &worklist);
  }
}
HBasicBlock* common_dominator = finder.Get();

//Step (2): iterate over the worklist to find sinking candidates
while (!worklist.empty()) {
  HInstruction* instruction = worklist.back();
  if (processed_instructions.IsBitSet(instruction->GetId())) {
    //The instruction has already been processed, continue. This
    //happens when the instruction is the input/user of multiple
    //instructions
    worklist.pop_back();
    continue;
  }
  bool all_users_in_post_dominated_blocks = true;
  bool can_move = true;
  //Check users of the instruction
  for (const HUseListNode<HInstruction*>& use :
       instruction->GetUses()) {
    HInstruction* user = use.GetUser();
    if (!post_dominated.IsBitSet(user->GetBlock()->GetBlockId()) &&
        !instructions_that_can_move.IsBitSet(user->GetId())) {
      all_users_in_post_dominated_blocks = false;
      //If we've already processed this user, or the user cannot be
      //moved, or is not dominating the post dominated blocks, bail
      //TODO(ngeoffray): The domination check is an approximation. We
      //should instead check if the dominated blocks post dominate the
      //user's block, but we do not have post dominance information
      //here
      if (processed_instructions.IsBitSet(user->GetId()) ||
          !IsInterestingInstruction(user) ||
          !user->GetBlock()->Dominates(common_dominator)) {
        can_move = false;
        break;
      }
    }
  }
```

```cpp
//Check environment users of the instruction. Some of these users
//require the instruction not to move
if (all_users_in_post_dominated_blocks) {
  for (const HUseListNode<HEnvironment*>& use :
          instruction->GetEnvUses()) {
    HEnvironment* environment = use.GetUser();
    HInstruction* user = environment->GetHolder();
    if (!post_dominated.IsBitSet(user->GetBlock()->GetBlockId())) {
      if (graph_->IsDebuggable() ||
          user->IsDeoptimize() ||
          user->CanThrowIntoCatchBlock() ||
          (user->IsSuspendCheck() && graph_->IsCompilingOsr())) {
        can_move = false;
        break;
      }
    }
  }
}
if (!can_move) {
  //Instruction cannot be moved, mark it as processed and remove it
  //from the work list
  processed_instructions.SetBit(instruction->GetId());
  worklist.pop_back();
} else if (all_users_in_post_dominated_blocks) {
  //Instruction is a candidate for being sunk. Mark it as such
  //remove it from the work list, and add its inputs to the work
  //list
  instructions_that_can_move.SetBit(instruction->GetId());
  move_in_order.push_back(instruction);
  processed_instructions.SetBit(instruction->GetId());
  worklist.pop_back();
  AddInputs(instruction, processed_instructions, post_dominated,
            &worklist);
  //Drop the environment use not in the list of post-dominated block
  //This is to help step (3) of this optimization, when we start
  //moving instructions closer to their use
  for (const HUseListNode<HEnvironment*>& use :
          instruction->GetEnvUses()) {
    HEnvironment* environment = use.GetUser();
    HInstruction* user = environment->GetHolder();
    if (!post_dominated.IsBitSet(user->GetBlock()->GetBlockId())) {
      environment->RemoveAsUserOfInput(use.GetIndex());
      environment->SetRawEnvAt(use.GetIndex(), nullptr);
    }
  }
} else {
```

```cpp
      //The information we have on the users was not enough to decide
      //whether the instruction could be moved
      //Add the users to the work list, and keep the instruction in the
      //work list to process it again once all users have been processed
      for (const HUseListNode<HInstruction*>& use :
              instruction->GetUses()) {
        AddInstruction(use.GetUser(), processed_instructions,
                  post_dominated, &worklist);
      }
    }
  }

  //Make sure we process instructions in dominated order. This is
  //required for heap stores
  std::sort(move_in_order.begin(), move_in_order.end(),
              [](HInstruction* a, HInstruction* b) {
    return b->StrictlyDominates(a);
  });

  //Step (3): Try to move sinking candidates
  for (HInstruction* instruction : move_in_order) {
    HInstruction* position = nullptr;
    if (instruction->IsArraySet()
            || instruction->IsInstanceFieldSet()
            || instruction->IsConstructorFence()) {
      if (!instructions_that_can_move.
                                IsBitSet(instruction->InputAt(0)->GetId())) {
        //A store can trivially move, but it can safely do so only if the
        //heap location it stores to can also move
        //TODO(ngeoffray): Handle allocation/store cycles by pruning
        //these instructions from the set and all their inputs
        continue;
      }
      //Find the position of the instruction we're storing into,
      //filtering out this store and all other stores to that
      //instruction
      position = FindIdealPosition(instruction->InputAt(0),
                                post_dominated, /* filter= */ true);

      //The position needs to be dominated by the store, in order for the
      //store to move there
      if (position == nullptr || !instruction->GetBlock()->Dominates(
          position->GetBlock())) {
        continue;
      }
    } else {
```

```cpp
      //Find the ideal position within the post dominated blocks
      position = FindIdealPosition(instruction, post_dominated);
      if (position == nullptr) {
        continue;
      }
    }
    //Bail if we could not find a position in the post dominated blocks
    //(for example, if there are multiple users whose common dominator
    //is not in the list of post dominated blocks)
    if (!post_dominated.IsBitSet(position->GetBlock()->GetBlockId())) {
      continue;
    }
    MaybeRecordStat(stats_, MethodCompilationStat::kInstructionSunk);
    instruction->MoveBefore(position, /* do_checks = */ false);
  }
}
```

SinkCodeToUncommonBranch()函数的主要执行过程可以分为3步：第1步，后序遍历end_block 的 post dominated，获取一个基本代码块子集合，并将该子集合添加到 worklist 中；第2步，遍历 worklist 寻找可以进行下沉的候选者；第3步，移动下沉候选者。需要专门说明的是第1步中的 end_block，并不是说它是最后一个基本代码块，而是 SinkCodeToUncommonBranch()函数的形参，它随着每次执行在不断地变化，只有这样才能保证解决所有的代码下沉的情况，而不只是解决最后一个基本代码块的情况。

本部分介绍了指令下沉的优化实现。此外，OptimizingCompiler 还有一些与机器无关的优化，具体如下：Hinliner、SideEffectsAnalysis、GVNOptimization、HSelectGenerator、LICM、HInductionVarAnalysis、LoadStoreAnalysis、LoadStoreElimination、CHAGuardOptimization 和 ConstructorFenceRedundancyElimination，在此不再一一展开进行介绍，读者可以根据需要查阅对应的代码。

4.7 硬件平台相关优化 pass 及其实现

前文介绍了 OptimizingCompiler 的与硬件平台无关的优化算法实例，在 OptimizingCompiler 中还有一部分与硬件平台相关的优化，这里将对这部分内容进行介绍。

与硬件平台相关的优化的主要实现位于 art/compiler/optimizing/optimizing_compiler.cc 中的 RunArchOptimizations()函数中。RunArchOptimizations()函数主要被同文件中的 TryCompileIntrinsic()函数和 5 个参数版本的 RunOptimizations()函数所调用，这两个函数都可以向上追溯到 OptimizingCompiler::Compile() 函数中，属于 OptimizingCompiler::Compile()函数执行过程中的不同分支。根据调用关系自上而下执行，OptimizingCompiler::Compile()函数会调用 TryCompileIntrinsic()函数和 TryCompile()

函数，TryCompile()函数会调用 RunBaselineOptimizations()函数或 5 个参数版本的 RunOptimizations()函数。

RunArchOptimizations()函数在执行的过程中，根据硬件架构的不同，采用宏去区分不同的硬件平台，然后为其配备不同的优化 pass 进行优化，代码如下：

```
//第 4 章/optimizing_compiler.cc
bool OptimizingCompiler::RunArchOptimizations(
    HGraph* graph, CodeGenerator* codegen,
    const DexCompilationUnit& dex_compilation_unit,
    PassObserver* pass_observer,
    VariableSizedHandleScope* handles) const {
  switch (codegen->GetCompilerOptions().GetInstructionSet()) {
#if defined(ART_ENABLE_CODEGEN_arm)
    case InstructionSet::kThumb2:
    case InstructionSet::kArm: {
      OptimizationDef arm_optimizations[] = {
        OptDef(OptimizationPass::kInstructionSimplifierArm),
        OptDef(OptimizationPass::kSideEffectsAnalysis),
        OptDef(OptimizationPass::kGlobalValueNumbering, "GVN$after_arch"),
        OptDef(OptimizationPass::kScheduling)
      };
      return RunOptimizations(graph,
                              codegen,
                              dex_compilation_unit,
                              pass_observer,
                              handles,
                              arm_optimizations);
    }
#endif
#ifdef ART_ENABLE_CODEGEN_arm64
    case InstructionSet::kArm64: {
      OptimizationDef arm64_optimizations[] = {
        OptDef(OptimizationPass::kInstructionSimplifierArm64),
        OptDef(OptimizationPass::kSideEffectsAnalysis),
        OptDef(OptimizationPass::kGlobalValueNumbering, "GVN$after_arch"),
        OptDef(OptimizationPass::kScheduling)
      };
      return RunOptimizations(graph,
                              codegen,
                              dex_compilation_unit,
                              pass_observer,
                              handles,
                              arm64_optimizations);
    }
#endif
```

```cpp
#ifdef ART_ENABLE_CODEGEN_mips
    case InstructionSet::kMips: {
      OptimizationDef mips_optimizations[] = {
        OptDef(OptimizationPass::kInstructionSimplifierMips),
        OptDef(OptimizationPass::kSideEffectsAnalysis),
        OptDef(OptimizationPass::kGlobalValueNumbering, "GVN$after_arch"),
        OptDef(OptimizationPass::kPcRelativeFixupsMips)
      };
      return RunOptimizations(graph,
                              codegen,
                              dex_compilation_unit,
                              pass_observer,
                              handles,
                              mips_optimizations);
    }
#endif
#ifdef ART_ENABLE_CODEGEN_mips64
    case InstructionSet::kMips64: {
      OptimizationDef mips64_optimizations[] = {
        OptDef(OptimizationPass::kSideEffectsAnalysis),
        OptDef(OptimizationPass::kGlobalValueNumbering, "GVN$after_arch")
      };
      return RunOptimizations(graph,
                              codegen,
                              dex_compilation_unit,
                              pass_observer,
                              handles,
                              mips64_optimizations);
    }
#endif
#ifdef ART_ENABLE_CODEGEN_x86
    case InstructionSet::kX86: {
      OptimizationDef x86_optimizations[] = {
        OptDef(OptimizationPass::kInstructionSimplifierX86),
        OptDef(OptimizationPass::kSideEffectsAnalysis),
        OptDef(OptimizationPass::kGlobalValueNumbering, "GVN$after_arch"),
        OptDef(OptimizationPass::kPcRelativeFixupsX86),
        OptDef(OptimizationPass::kX86MemoryOperandGeneration)
      };
      return RunOptimizations(graph,
                              codegen,
                              dex_compilation_unit,
                              pass_observer,
                              handles,
                              x86_optimizations);
    }
```

```
# endif
# ifdef ART_ENABLE_CODEGEN_x86_64
    case InstructionSet::kX86_64: {
      OptimizationDef x86_64_optimizations[] = {
        OptDef(OptimizationPass::kInstructionSimplifierX86_64),
        OptDef(OptimizationPass::kSideEffectsAnalysis),
        OptDef(OptimizationPass::kGlobalValueNumbering, "GVN$after_arch"),
        OptDef(OptimizationPass::kX86MemoryOperandGeneration)
      };
      return RunOptimizations(graph,
                              codegen,
                              dex_compilation_unit,
                              pass_observer,
                              handles,
                              x86_64_optimizations);
    }
# endif
    default:
      return false;
  }
}
```

其中采用 switch-case 的结构，结合与各个硬件平台相关的宏，构成了整个函数的主体结构。在这里支持了 arm、arm64、mips、mips64、x86 和 x86_64 共 6 种架构，为每种架构定制了 pass 组合构成硬件平台相关的优化。需要特别说明的是，这里在硬件相关优化的 pass 组合中，并不是每个 pass 都是硬件相关的 pass。以 x86_64 架构为例，组织了 kInstructionSimplifierX86_64、kSideEffectsAnalysis、kGlobalValueNumbering 和 kX86MemoryOperandGeneration 共 4 个 pass，只有第 1 个和第 4 个 pass 是与硬件平台相关的 pass，第 2 个和第 3 个 pass 依然是与硬件平台无关的优化 pass。甚至在 mips64 硬件平台下，所有的两个 pass 都是与硬件无关优化 pass，但是也不妨碍其整体作为 mips64 硬件平台的一个优化方案存在。

所以，与硬件平台相关的优化，可以有部分的硬件平台相关优化 pass 参与，也可以完全没有硬件平台优化 pass 参与。除了 RunArchOptimizations() 函数中有与硬件平台相关的优化，在同文件的 RunBaselineOptimizations() 函数中，也有硬件平台的相关优化存在。

RunBaselineOptimizations() 函数是 TryCompile() 函数在执行的时候会选择调用的两个函数之一，另外一个是有 5 个参数版本的 RunOptimizations() 函数，它们之间的关系在上文已经介绍过。RunBaselineOptimizations() 函数中也会根据硬件平台为其准备与硬件平台相关的优化，代码如下：

```
//第 4 章/optimizing_compiler.cc
bool OptimizingCompiler::RunBaselineOptimizations(
    HGraph* graph, CodeGenerator* codegen,
```

```cpp
            const DexCompilationUnit& dex_compilation_unit,
            PassObserver* pass_observer,
            VariableSizedHandleScope* handles) const {
    switch (codegen->GetCompilerOptions().GetInstructionSet()) {
#ifdef ART_ENABLE_CODEGEN_mips
        case InstructionSet::kMips: {
            OptimizationDef mips_optimizations[] = {
                OptDef(OptimizationPass::kPcRelativeFixupsMips)
            };
            return RunOptimizations(graph,
                                    codegen,
                                    dex_compilation_unit,
                                    pass_observer,
                                    handles,
                                    mips_optimizations);
        }
#endif
#ifdef ART_ENABLE_CODEGEN_x86
        case InstructionSet::kX86: {
            OptimizationDef x86_optimizations[] = {
                OptDef(OptimizationPass::kPcRelativeFixupsX86),
            };
            return RunOptimizations(graph,
                                    codegen,
                                    dex_compilation_unit,
                                    pass_observer,
                                    handles,
                                    x86_optimizations);
        }
#endif
        default:
            UNUSED(graph);
            UNUSED(codegen);
            UNUSED(dex_compilation_unit);
            UNUSED(pass_observer);
            UNUSED(handles);
            return false;
    }
}
```

RunBaselineOptimizations()函数和 RunArchOptimizations()函数不同,这里只为 mips 和 x86 准备了相关的优化 pass 组合,并且每个组合之内只有一个硬件相关的 pass。为 mips 硬件平台准备了 kPcRelativeFixupsMips pass,为 x86 平台准备了 kPcRelativeFixupsX86 pass。

综合 RunArchOptimizations()函数和 RunBaselineOptimizations()函数中的所有与硬件平台相关的优化 pass,如 kInstructionSimplifierArm、kInstructionSimplifierArm64、 kInstructionSimplifierMips、kPcRelativeFixupsMips、kInstructionSimplifierX86、kPcRelativeFixupsX86、kX86MemoryOperandGeneration 和 kInstructionSimplifierX86_64，共 8 个 pass；其中，还用到了一个之前没有使用过的与硬件平台无关的 kScheduling pass。

kInstructionSimplifierArm pass 对应的实现类为 InstructionSimplifierArm，InstructionSimplifierArm 类的实现位于 art/compiler/optimizing/instruction_simplifier_arm.h 和 art/compiler/optimizing/instruction_simplifier_arm.cc 中。kInstructionSimplifierArm64 对应的实现类为 InstructionSimplifierArm64，InstructionSimplifierArm64 类的实现位于 art/compiler/optimizing/instruction_simplifier_arm64.h 和 art/compiler/optimizing/instruction_simplifier_arm64.cc 中。

kInstructionSimplifierMips pass 对应的实现类为 InstructionSimplifierMips，InstructionSimplifierMips 类的实现位于 art/compiler/optimizing/instruction_simplifier_mips.h 和 art/compiler/optimizing/instruction_simplifier_mips.cc 中。kPcRelativeFixupsMips pass 对应的实现类为 mips::PcRelativeFixups，mips::PcRelativeFixups 类的实现位于 art/compiler/optimizing/pc_relative_fixups_mips.h 和 art/compiler/optimizing/pc_relative_fixups_mips.cc 中。

kInstructionSimplifierX86 pass 对应的实现类为 InstructionSimplifierX86，InstructionSimplifierX86 类的实现位于 art/compiler/optimizing/instruction_simplifier_x86.h 和 art/compiler/optimizing/instruction_simplifier_x86.cc 中。kPcRelativeFixupsX86 pass 对应的实现类为 x86::PcRelativeFixups，x86::PcRelativeFixups 类的实现位于 art/compiler/optimizing/pc_relative_fixups_x86.h 和 art/compiler/optimizing/pc_relative_fixups_x86.cc 中。kX86MemoryOperandGeneration pass 对应的实现类为 X86MemoryOperandGeneration，X86MemoryOperandGeneration 类的实现位于 art/compiler/optimizing/x86_memory_gen.h 和 art/compiler/optimizing/x86_memory_gen.cc 中。kInstructionSimplifierX86_64 pass 对应的实现类为 InstructionSimplifierX86_64，InstructionSimplifierX86_64 类的实现位于 art/compiler/optimizing/instruction_simplifier_x86_64.h 和 art/compiler/optimizing/instruction_simplifier_x86_64.cc 中。

kScheduling pass 对应的实现类为 HInstructionScheduling，HInstructionScheduling 的实现位于 art/compiler/optimizing/scheduler.h 和 art/compiler/optimizing/scheduler.cc 中。

上述 9 个在硬件相关优化中使用的 pass 及其实现都已经进行了简单的介绍，如果有需要，则可以继续深入分析对应的文件。

本部分对于硬件相关平台优化的整体情况及其具体实现进行了介绍，读者可以根据实际所使用的具体硬件平台来分析其所对应的优化内容。

4.8　小结

本章介绍了优化 pass 所对应的实现类，其中优化 pass 可以分为与平台无关的优化 pass 和与平台相关的优化 pass，这些 pass 所对应的实现类都是 HOptimization 的子类。

需要注意的是，有些 pass 及其实现没有在优化列表中直接调用，而是被其他的优化 pass 调用，例如：SsaRedundantPhiElimination 作为 HOptimization 的子类，它也是一个优化 pass 的实现类，但是它是被 HDeadCodeElimination 调用的，并没有在优化 pass 列表中被直接调用。

在 HOptimization 的子类构成的优化体系之外，还有一个辅助体系，它是 HGraphVisitor 及其子类构成的遍历访问体系，这个体系是配合 HOptimization 的子类构成的优化体系，能更好地访问节点和进行优化，但是，并不是所有优化的实现类都有配合的 XXXXVisitor 子类有。

本部分在介绍了优化框架的基础之上，对部分具有代表性的 pass 进行了介绍，旨在描述整体优化框架的基础之上，让读者能对具体实现也有一定的了解。

第 5 章 ART 启动分析

前文介绍了 AOT 编译和 JIT 编译所使用的 OptimizingCompiler 的相关内容,这些内容也是 dex2oat 工具的核心,dex2oat 也属于 ART 的范围。在接下来的内容中,将对 ART 的运行时部分进行介绍,主要指的是 ART 的启动、运行和内存管理等内容。

ART 的启动环节,包含了 ART 中所有重要部分的初始化和启动,理解这部分内容是理解整个 ART 的重要基础。本章将分析 ART 启动中多个部分的启动和实现,为读者展示 ART 启动中的关键流程。

5.1 ART 启动中的虚拟机启动一

ART 是 Android 的运行时,其内部还包含了一个虚拟机,这个虚拟机主要用于指令的执行。本部分内容将对 ART 启动中虚拟机的启动进行分析。

ART 是由 Zygote 进程所启动的,Zygote 所对应的启动代码位于 frameworks/base/cmds/app_process/app_main.cpp 中的 main()函数。这里的 main()函数通过构建一个新的 AppRuntime 对象,然后调用其 start()函数来启动运行时,代码如下:

```cpp
//第 5 章/app_main.cpp
int main(int argc, char* const argv[])
{
    if (!LOG_NDEBUG) {
      String8 argv_String;
      for (int i = 0; i < argc; ++i) {
        argv_String.append("\"");
        argv_String.append(argv[i]);
        argv_String.append("\" ");
      }
      ALOGV("app_process main with argv: %s", argv_String.string());
    }

    AppRuntime Runtime(argv[0], computeArgBlockSize(argc, argv));
    //Process command line arguments
```

```
//ignore argv[0]
argc--;
argv++;

//Everything up to '--' or first non '-' arg goes to the vm
//
//The first argument after the VM args is the "parent dir", which
//is currently unused
//
//After the parent dir, we expect one or more the following internal
//arguments
//
// -- zygote : Start in zygote mode
// -- start-system-server : Start the system server
// -- application : Start in application (stand alone, non zygote) mode
// -- nice-name : The nice name for this process
//
//For non zygote starts, these arguments will be followed by
//the main class name. All remaining arguments are passed to
//the main method of this class
//
//For zygote starts, all remaining arguments are passed to the
//zygote main function
//
//Note that we must copy argument string values since we will
//rewrite the entire argument block when we apply the nice name to argv0
//
//As an exception to the above rule, anything in "spaced commands"
//goes to the vm even though it has a space in it
const char* spaced_commands[] = { "-cp", "-classpath" };
//Allow "spaced commands" to be succeeded by exactly 1 argument
//(regardless of -s)
bool known_command = false;

int i;
for (i = 0; i < argc; i++) {
    if (known_command == true) {
        Runtime.addOption(strdup(argv[i]));
        //The static analyzer gets upset that we don't ever free the
        //above string. Since the allocation is from main, leaking it
        //doesn't seem problematic. NOLINTNEXTLINE
        ALOGV("app_process main add known option '%s'", argv[i]);
        known_command = false;
        continue;
    }
```

```cpp
            for (int j = 0;
                j < static_cast<int>(sizeof(spaced_commands) /
                sizeof(spaced_commands[0])); ++j) {
              if (strcmp(argv[i], spaced_commands[j]) == 0) {
                known_command = true;
                ALOGV("app_process main found known command '%s'", argv[i]);
              }
            }

            if (argv[i][0] != '-') {
                break;
            }
            if (argv[i][1] == '-' && argv[i][2] == 0) {
                ++i; //Skip --.
                break;
            }

            Runtime.addOption(strdup(argv[i]));
            //The static analyzer gets upset that we don't ever free the
            //above string. Since the allocation is from main, leaking it
            //doesn't seem problematic. NOLINTNEXTLINE
            ALOGV("app_process main add option '%s'", argv[i]);
    }

    //Parse Runtime arguments. Stop at first unrecognized option
    bool zygote = false;
    bool startSystemServer = false;
    bool application = false;
    String8 niceName;
    String8 className;

    ++i; //Skip unused "parent dir" argument
    while (i < argc) {
        const char* arg = argv[i++];
        if (strcmp(arg, "--zygote") == 0) {
            zygote = true;
            niceName = ZYGOTE_NICE_NAME;
        } else if (strcmp(arg, "--start-system-server") == 0) {
            startSystemServer = true;
        } else if (strcmp(arg, "--application") == 0) {
            application = true;
        } else if (strncmp(arg, "--nice-name=", 12) == 0) {
            niceName.setTo(arg + 12);
        } else if (strncmp(arg, "--", 2) != 0) {
            className.setTo(arg);
            break;
```

```cpp
        } else {
            --i;
            break;
        }
    }

    Vector<String8> args;
    if (!className.isEmpty()) {
        //We're not in zygote mode, the only argument we need to pass
        //to RuntimeInit is the application argument
        //
        //The Remainder of args get passed to startup class main(). Make
        //copies of them before we overwrite them with the process name
        args.add(application ? String8("application") : String8("tool"));
        Runtime.setClassNameAndArgs(className, argc - i, argv + i);

        if (!LOG_NDEBug) {
          String8 restOfArgs;
          char * const * argv_new = argv + i;
          int argc_new = argc - i;
          for (int k = 0; k < argc_new; ++k) {
            restOfArgs.append("\"");
            restOfArgs.append(argv_new[k]);
            restOfArgs.append("\" ");
          }
          ALOGV("Class name = %s, args = %s", className.string(),
                restOfArgs.string());
        }
    } else {
        //We're in zygote mode
        maybeCreateDalvikCache();

        if (startSystemServer) {
            args.add(String8("start-system-server"));
        }

        char prop[PROP_VALUE_MAX];
        if (property_get(ABI_LIST_PROPERTY, prop, NULL) == 0) {
            LOG_ALWAYS_FATAL("app_process: Unable to determine ABI list
                            from property %s.", ABI_LIST_PROPERTY);
            return 11;
        }

        String8 abiFlag("--abi-list=");
        abiFlag.append(prop);
        args.add(abiFlag);
```

```
            //In zygote mode, pass all remaining arguments to the zygote
            //main() method
            for (; i < argc; ++i) {
                args.add(String8(argv[i]));
            }
        }

        if (!niceName.isEmpty()) {
            Runtime.setArgv0(niceName.string(), true /* setProcName */);
        }

        if (zygote) {
            Runtime.start("com.android.internal.os.ZygoteInit", args, zygote);
        } else if (className) {
            Runtime.start("com.android.internal.os.RuntimeInit", args, zygote);
        } else {
            fprintf(stderr, "Error: no class name or -- zygote supplied.\n");
            app_usage();
            LOG_ALWAYS_FATAL("app_process: no class name or -- zygote supplied.");
        }
    }
```

AppRuntime 类继承于 AndroidRuntime 类,它的定义和实现位于 frameworks/base/cmds/app_process/app_main.cpp 中,它并没有重新定义 start()函数,所以它对 start()函数的调用应该是调用其父类 AndroidRuntime 的 start()函数。AndroidRuntime 类的定义和实现位于 frameworks/base/core/jni/include/android_Runtime/ AndroidRuntime.h 和 frameworks/base/core/jni/AndroidRuntime.cpp 中。AndroidRuntime 的 start()函数位于 AndroidRuntime.cpp 中,它用于启动 ART,其核心内容是通过调用 JniInvocation 的 Init()函数准备相关内容,再通过 AndroidRuntime 的 startVm()函数启动虚拟机。此处,我们先聚焦于虚拟机的启动。其实,ART 所覆盖的范围要大于虚拟机,ART 指的是整个运行时体系,包含了虚拟机、编译器、dex2oat 工具等众多内容。AndroidRuntime::start()函数的代码如下:

```
//第 5 章/AndroidRuntime.cpp
/*
 * Start the Android Runtime.  This involves starting the virtual machine
 * and calling the "static void main(String[] args)" method in the class
 * named by "className".
 *
 * Passes the main function two arguments, the class name and the  * specified options string.
 */
```

```cpp
void AndroidRuntime::start(
    const char* className, const Vector<String8>& options, bool zygote) {
    ALOGD(">>>>>> START %s uid %d <<<<<<\n",
            className != NULL ? className : "(unknown)", getuid());

    static const String8 startSystemServer("start-system-server");

    /*
     * 'startSystemServer == true' means Runtime is obsolete and not run
     * from init.rc anymore, so we print out the boot start event here.
     */
    for (size_t i = 0; i < options.size(); ++i) {
        if (options[i] == startSystemServer) {
            /* track our progress through the boot sequence */
            const int LOG_BOOT_PROGRESS_START = 3000;
            LOG_EVENT_LONG(LOG_BOOT_PROGRESS_START,
                           ns2ms(systemTime(SYSTEM_TIME_MONOTONIC)));
        }
    }

    const char* rootDir = getenv("ANDROID_ROOT");
    if (rootDir == NULL) {
        rootDir = "/system";
        if (!hasDir("/system")) {
            LOG_FATAL("No root directory specified, "
                      "and /system does not exist.");
            return;
        }
        setenv("ANDROID_ROOT", rootDir, 1);
    }

    const char* RuntimeRootDir = getenv("ANDROID_RUNTIME_ROOT");
    if (RuntimeRootDir == NULL) {
        LOG_FATAL("No Runtime directory specified with "
                  "ANDROID_RUNTIME_ROOT environment variable.");
        return;
    }

    const char* tzdataRootDir = getenv("ANDROID_TZDATA_ROOT");
    if (tzdataRootDir == NULL) {
        LOG_FATAL("No tz data directory specified with "
                  "ANDROID_TZDATA_ROOT environment variable.");
        return;
    }

    //const char* KernelHack = getenv("LD_ASSUME_KERNEL");
```

```
//ALOGD("Found LD_ASSUME_KERNEL = '%s'\n", KernelHack);

/* start the virtual machine */
JniInvocation jni_invocation;
jni_invocation.Init(NULL);
JNIEnv* env;
if (startVm(&mJavaVM, &env, zygote) != 0) {
    return;
}
onVmCreated(env);

/*
 * Register android functions.
 */
if (startReg(env) < 0) {
    ALOGE("Unable to register all android natives\n");
    return;
}

/*
 * We want to call main() with a String array with arguments in it.
 * At present we have two arguments, the class name and an option
 * string. Create an array to hold them.
 */
jclass stringClass;
jobjectArray strArray;
jstring classNameStr;

stringClass = env->FindClass("java/lang/String");
assert(stringClass != NULL);
strArray = env->NewObjectArray(options.size() + 1, stringClass,
                               NULL);
assert(strArray != NULL);
classNameStr = env->NewStringUTF(className);
assert(classNameStr != NULL);
env->SetObjectArrayElement(strArray, 0, classNameStr);

for (size_t i = 0; i < options.size(); ++i) {
    jstring optionsStr =
        env->NewStringUTF(options.itemAt(i).string());
    assert(optionsStr != NULL);
    env->SetObjectArrayElement(strArray, i + 1, optionsStr);
}

/*
 * Start VM. This thread becomes the main thread of the VM, and will
```

```
     * not return until the VM exits.
     */
    char* slashClassName = toSlashClassName(
        className != NULL ? className : "");
    jclass startClass = env->FindClass(slashClassName);
    if (startClass == NULL) {
        ALOGE("JavaVM unable to locate class '%s'\n", slashClassName);
        /* keep going */
    } else {
        jmethodID startMeth = env->GetStaticMethodID(startClass, "main",
            "([Ljava/lang/String;)V");
        if (startMeth == NULL) {
            ALOGE("JavaVM unable to find main() in '%s'\n", className);
            /* keep going */
        } else {
            env->CallStaticVoidMethod(startClass, startMeth, strArray);

#if 0
            if (env->ExceptionCheck())
                threadExitUncaughtException(env);
#endif
        }
    }
    free(slashClassName);

    ALOGD("Shutting down VM\n");
    if (mJavaVM->DetachCurrentThread() != JNI_OK)
        ALOGW("Warning: unable to detach main thread\n");
    if (mJavaVM->DestroyJavaVM() != 0)
        ALOGW("Warning: VM did not shut down cleanly\n");
}
```

JniInvocation 的 Init()函数的实现位于 libnativehelper/include/nativehelper/JniInvocation.h 中，主要用于初始化 Jni 调用的 API，代码如下：

```
//第5章/JniInvocation.h
    //Initialize JNI invocation API. library should specifiy a valid
    //shared library for opening via dlopen providing a JNI invocation
    //implementation, or null to allow defaulting via
    //persist.sys.dalvik.vm.lib
    bool Init(const char* library) {
        return JniInvocationInit(impl_, library) != 0;
    }
```

其中 impl_是一个 JniInvocationImpl 类型的指针，并且在构造函数中进行了初始化，其

指向一个新建的 JniInvocationImpl 对象,代码如下:

```
//第 5 章/JniInvocation.h
JniInvocationImpl* impl_;

  JniInvocation() {
    impl_ = JniInvocationCreate();
  }

MODULE_API JniInvocationImpl* JniInvocationCreate() {
  return new JniInvocationImpl();
}
```

JniInvocationCreate()函数的实现位于 libnativehelper/JniInvocation.cpp 中。在明确了 impl_之后,libnativehelper/JniInvocation.cpp 中的 JniInvocationInit()函数将对 library 初始化,转化为使用 impl_ 的 Init()函数对 library 进行初始化,代码如下:

```
//第 5 章/JniInvocation.cpp
MODULE_API int JniInvocationInit(JniInvocationImpl* instance, const char*
library) {
  return instance->Init(library) ? 1 : 0;
}
```

其中的 instance 是 JniInvocationImpl 类型的指针,JniInvocation 的 Init()函数最终被转化为 libnativehelper/JniInvocation.cpp 中的 JniInvocationImpl::Init() 函数。JniInvocationImpl::Init()函数的最重要的工作是通过 GetLibrary()函数获取库,并且通过 OpenLibrary()函数打开库,然后通过 FindSymbol 将 JNI_GetDefaultJavaVMInitArgs()函数挂载到 JNI_GetDefaultJavaVMInitArgs_,将 JNI_CreateJavaVM()函数挂载到 JNI_CreateJavaVM_,将 JNI_GetCreatedJavaVMs()函数挂载到 JNI_GetCreatedJavaVMs_。JniInvocationImpl::Init()函数的具体实现如下:

```
//第 5 章/JniInvocation.cpp
bool JniInvocationImpl::Init(const char* library) {
#ifdef __ANDROID__
  char buffer[PROP_VALUE_MAX];
#else
  char* buffer = NULL;
#endif
  library = GetLibrary(library, buffer);
  handle_ = OpenLibrary(library);
  if (handle_ == NULL) {
    if (strcmp(library, kLibraryFallback) == 0) {
      //Nothing else to try
```

```cpp
      ALOGE("Failed to dlopen %s: %s", library, GetError().c_str());
      return false;
    }
    //Note that this is enough to get something like the zygote
    //running, we can't property_set here to fix this for the future
    //because we are root and not the system user. See
    //RuntimeInit.commonInit for where we fix up the property to
    //avoid future fallbacks. http://b/11463182
    ALOGW("Falling back from %s to %s after dlopen error: %s",
          library, kLibraryFallback, GetError().c_str());
    library = kLibraryFallback;
    handle_ = OpenLibrary(library);
    if (handle_ == NULL) {
      ALOGE("Failed to dlopen %s: %s", library, GetError().c_str());
      return false;
    }
  }
  if (!FindSymbol(reinterpret_cast<FUNC_POINTER*>(
      &JNI_GetDefaultJavaVMInitArgs_), "JNI_GetDefaultJavaVMInitArgs")) {
    return false;
  }
  if (!FindSymbol(reinterpret_cast<FUNC_POINTER*>(&JNI_CreateJavaVM_),
      "JNI_CreateJavaVM")) {
    return false;
  }
  if (!FindSymbol(reinterpret_cast<FUNC_POINTER*>(
      &JNI_GetCreatedJavaVMs_), "JNI_GetCreatedJavaVMs")) {
    return false;
  }
  return true;
}
```

启动 ART 时默认传递的 library 值是 NULL, 在 GetLibrary 的时候, 如果 library 为空, 则会将 libart.so 传递给 library, 所以是从 libart.so 库中寻找 JNI_GetDefaultJava-VMInitArgs() 函数、JNI_CreateJavaVM() 函数和 JNI_GetCreatedJavaVMs() 函数, 这 3 个函数都位于 art/Runtime/jni/java_vm_ext.cc 中, 这样就进入了 ART 的源码目录。

Init 过程结束之后, 进行的是启动虚拟机的过程。AndroidRuntime 的 startVM() 函数的实现位于 frameworks/base/core/jni/AndroidRuntime.cpp 中, 通过 JNI_CreateJavaVM() 函数启动 Dalvik 虚拟机, 代码如下:

```cpp
//第5章/AndroidRuntime.cpp
/*
 * Start the Dalvik Virtual Machine.
 *
```

```
 * Various arguments, most determined by system properties, are passed
 * in.
 * The "mOptions" vector is updated.
 *
 * CAUTION: when adding options in here, be careful not to put the
 * char buffer inside a nested scope. Adding the buffer to the
 * options using mOptions.add() does not copy the buffer, so if the
 * buffer goes out of scope the option may be overwritten. It's best
 * to put the buffer at the top of the function so that it is more
 * unlikely that someone will surround it in a scope at a later time
 * and thus introduce a bug.
 *
 * Returns 0 on success.
 */
int AndroidRuntime::startVm(JavaVM** pJavaVM, JNIEnv** pEnv, bool zygote)
{
    ...
    /*
     * Initialize the VM.
     *
     * The JavaVM* is essentially per-process, and the JNIEnv* is per-
     * thread.
     * If this call succeeds, the VM is ready, and we can start issuing
     * JNI calls.
     */
    if (JNI_CreateJavaVM(pJavaVM, pEnv, &initArgs) < 0) {
        ALOGE("JNI_CreateJavaVM failed\n");
        return -1;
    }

    return 0;
}
```

其中,这里的 JNI_CreateJavaVM() 函数指的是 libnativehelper/JniInvocation.cpp 中的 JNI_CreateJavaVM() 函数,JniInvocation.cpp 中的 JNI_CreateJavaVM() 函数将调用转到 JniInvocationImpl::JNI_CreateJavaVM() 函数中,最终调用了 JNI_CreateJavaVM_() 函数,即在 JniInvocationImpl::Init() 环节挂载的 art/Runtime/jni/java_vm_ext.cc 中的 JNI_CreateJavaVM() 函数。实现上述过程的代码如下:

```
//第 5 章/JniInvocation.cpp
MODULE_API jint JNI_CreateJavaVM(
    JavaVM** p_vm, JNIEnv** p_env, void* vm_args) {
  //Ensure any cached heap objects from previous VM instances are
  //invalidated. There is no notification here that a VM is destroyed.
```

```
//These cached objects limit us to one VM instance per process
  JniConstants::Uninitialize();
  return JniInvocationImpl::GetJniInvocation().JNI_CreateJavaVM(
      p_vm, p_env, vm_args);
}

jint JniInvocationImpl::JNI_CreateJavaVM(
    JavaVM** p_vm, JNIEnv** p_env, void* vm_args) {
  return JNI_CreateJavaVM_(p_vm, p_env, vm_args);
}
```

到这个阶段，Dalivk 虚拟机就启动成功了。ART 内部关于虚拟机启动的相关代码可以从 art/Runtime/jni/java_vm_ext.cc 中的 JNI_CreateJavaVM() 函数开始，继续向下跟踪，这会在后续部分进行介绍。

本部分介绍了 ART 启动过程中虚拟机启动的开头部分内容，后续内容将在 5.2 节继续展开介绍。

5.2　ART 启动中的虚拟机启动二

前文在分析虚拟机的启动流程中分析了 art/Runtime/jni/java_vm_ext.cc 中的 JNI_CreateJavaVM() 函数。本部分内容将从 java_vm_ext.cc 中的 JNI_CreateJavaVM() 函数继续向下跟踪，为读者展示 art 源码目录中的虚拟机启动的流程和操作。

java_vm_ext.cc 中的 JNI_CreateJavaVM() 函数通过 Runtime::Create() 函数构建一个新的 Runtime，然后通过 Runtime 调用其 Start() 函数，代码如下：

```
//第 5 章/java_vm_ext.cc
//JNI Invocation interface

extern "C" jint JNI_CreateJavaVM(
    JavaVM** p_vm, JNIEnv** p_env, void* vm_args) {
  ScopedTrace trace(__FUNCTION__);
  const JavaVMInitArgs* args = static_cast<JavaVMInitArgs*>(vm_args);
  if (JavaVMExt::IsBadJniVersion(args->version)) {
    LOG(ERROR) << "Bad JNI version passed to CreateJavaVM: "
               << args->version;
    return JNI_EVERSION;
  }
  RuntimeOptions options;
  for (int i = 0; i < args->nOptions; ++i) {
    JavaVMOption* option = &args->options[i];
    options.push_back(std::make_pair(std::string(option->optionString),
        option->extraInfo));
```

```
    }
    bool ignore_unrecognized = args->ignoreUnrecognized;
    if (!Runtime::Create(options, ignore_unrecognized)) {
      return JNI_ERR;
    }

    //Initialize native loader. This step makes sure we have
    //everything set up before we start using JNI
    android::InitializeNativeLoader();

    Runtime* Runtime = Runtime::Current();
    bool started = Runtime->Start();
    if (!started) {
      delete Thread::Current()->GetJniEnv();
      delete Runtime->GetJavaVM();
      LOG(WARNING) << "CreateJavaVM failed";
      return JNI_ERR;
    }

    *p_env = Thread::Current()->GetJniEnv();
    *p_vm = Runtime->GetJavaVM();
    return JNI_OK;
}
```

Runtime 类的实现位于 art/Runtime/Runtime.cc 中，它的 Create()函数有两个实现，其中两个参数版本的实现最终被转换为对 Runtime 类的 ParseOptions()函数和 1 个参数版本实现的调用，1 个参数版本的实现最终调用了 Init()函数，其中 Init()函数主要用于对 Runtime 进行初始化。Create()函数的两个实现如下：

```
//第 5 章/Runtime.cc
bool Runtime::Create(RuntimeArgumentMap&& Runtime_options) {
  //TODO: acquire a static mutex on Runtime to avoid racing
  if (Runtime::instance_ != nullptr) {
    return false;
  }
  instance_ = new Runtime;
  Locks::SetClientCallback(IsSafeToCallAbort);
  if (!instance_->Init(std::move(Runtime_options))) {
    //TODO: Currently deleting the instance will abort the Runtime on
    //destruction. Now This will leak memory, instead. Fix the
    //destructor. b/19100793
    //delete instance_
    instance_ = nullptr;
    return false;
```

```
    }
    return true;
}

bool Runtime::Create(
        const RuntimeOptions& raw_options, bool ignore_unrecognized) {
    RuntimeArgumentMap Runtime_options;
    return ParseOptions(raw_options, ignore_unrecognized,
        &Runtime_options) && Create(std::move(Runtime_options));
}
```

JNI_CreateJavaVM()函数中,在 Runtime 的 Create()函数执行之后,还调用了 Start()函数,也就是完成了 Runtime 的创建和初始化之后,进行了 Runtime 对象的启动。Start()函数的代码如下:

```
//第5章/Runtime.cc
bool Runtime::Start() {
    VLOG(startup) << "Runtime::Start entering";

    CHECK(!no_sig_chain_) <<
        "A started Runtime should have sig chain enabled";

    //If a debug host build, disable ptrace restriction for debugging and
    //test timeout thread dump
    //Only 64 - bit as prctl() may fail in 32 bit userspace on a 64 - bit Kernel
#if defined(__Linux__) && !defined(ART_TARGET_ANDROID) && defined(__x86_64__)
    if (kIsDebugBuild) {
        CHECK_EQ(prctl(PR_SET_PTRACER, PR_SET_PTRACER_ANY), 0);
    }
#endif

    //Restore main thread state to kNative as expected by native code
    Thread * self = Thread::Current();

    self -> TransitionFromRunnableToSuspended(kNative);

    DoAndMaybeSwitchInterpreter([ = ](){ started_ = true; });

    if (!IsImageDex2OatEnabled() || !GetHeap() -> HasBootImageSpace()) {
        ScopedObjectAccess soa(self);
        StackHandleScope < 2 > hs(soa.Self());

        ObjPtr < mirror::ObjectArray < mirror::Class >> class_roots =
            GetClassLinker() -> GetClassRoots();
        auto class_class(hs.NewHandle < mirror::Class >(
```

```cpp
            GetClassRoot<mirror::Class>(class_roots)));
    auto field_class(hs.NewHandle<mirror::Class>(
        GetClassRoot<mirror::Field>(class_roots)));

    class_linker_->EnsureInitialized(soa.Self(), class_class, true, true);
    self->AssertNoPendingException();
    //Field class is needed for register_java_net_InetAddress in
    //libcore, b/28153851
    class_linker_->EnsureInitialized(soa.Self(), field_class, true, true);
    self->AssertNoPendingException();
}

//InitNativeMethods needs to be after started_ so that the classes
//it touches will have methods linked to the oat file if necessary
{
    ScopedTrace trace2("InitNativeMethods");
    InitNativeMethods();
}

//IntializeIntrinsics needs to be called after the
//WellKnownClasses::Init in InitNativeMethods
//because in checking the invocation types of intrinsic methods
//ArtMethod::GetInvokeType() needs the SignaturePolymorphic annotation
//class which is initialized in WellKnownClasses::Init
InitializeIntrinsics();

//Initialize well known thread group values that may be accessed
//threads while attaching
InitThreadGroups(self);

Thread::FinishStartup();

//Create the JIT either if we have to use JIT compilation or save
//profiling info. This is done after FinishStartup as the JIT pool
//needs Java thread peers, which require the main ThreadGroup to
//exist
//
//TODO(calin): We use the JIT class as a proxy for JIT compilation and
//for recoding profiles. Maybe we should consider changing the name to
//be more clear it's not only about compiling. b/28295073
if (jit_options_->UseJitCompilation() || jit_options_->GetSaveProfilingInfo()) {
    //Try to load compiler pre zygote to reduce PSS. b/27744947
    std::string error_msg;
    if (!jit::Jit::LoadCompilerLibrary(&error_msg)) {
        LOG(WARNING) << "Failed to load JIT compiler with error "
                    << error_msg;
    }
```

```cpp
    CreateJitCodeCache(/*rwx_memory_allowed=*/true);
    CreateJit();
}

//Send the start phase event. We have to wait till here as this is
//when the main thread peer has just been generated, important root
//clinits have been run and JNI is completely functional
{
    ScopedObjectAccess soa(self);
    callbacks_->NextRuntimePhase(
        RuntimePhaseCallback::RuntimePhase::kStart);
}

system_class_loader_ = CreateSystemClassLoader(this);

if (!is_zygote_) {
    if (is_native_bridge_loaded_) {
        PreInitializeNativeBridge(".");
    }
    NativeBridgeAction action = force_native_bridge_
        ? NativeBridgeAction::kInitialize
        : NativeBridgeAction::kUnload;
    InitNonZygoteOrPostFork(self->GetJniEnv(),
        /*is_system_server=*/false, action,
        GetInstructionSetString(kRuntimeISA));
}

StartDaemonThreads();

//Make sure the environment is still clean (no lingering local refs
//from starting daemon threads)
{
    ScopedObjectAccess soa(self);
    self->GetJniEnv()->AssertLocalsEmpty();
}

//Send the initialized phase event. Send it after starting the Daemon
//threads so that agents cannot delay the daemon threads from starting
//forever
{
    ScopedObjectAccess soa(self);
    callbacks_->NextRuntimePhase(
        RuntimePhaseCallback::RuntimePhase::kInit);
}

{
```

```cpp
    ScopedObjectAccess soa(self);
    self->GetJniEnv()->AssertLocalsEmpty();
}

VLOG(startup) << "Runtime::Start exiting";
finished_starting_ = true;

if (trace_config_.get() != nullptr && trace_config_->trace_file != "") {
    ScopedThreadStateChange tsc(self, kWaitingForMethodTracingStart);
    Trace::Start(trace_config_->trace_file.c_str(),
                 static_cast<int>(trace_config_->trace_file_size),
                 0,
                 trace_config_->trace_output_mode,
                 trace_config_->trace_mode,
                 0);
}

//In case we have a profile path passed as a command line argument
//register the current class path for profiling now. Note that we
//cannot do this before we create the JIT and having it here is the
//most convenient way. This is used when testing profiles with
//dalvikvm command as there is no framework to register the dex files
//for profiling
if (jit_.get() != nullptr && jit_options_->GetSaveProfilingInfo() &&
    !jit_options_->GetProfileSaverOptions().GetProfilePath().empty()) {
    std::vector<std::string> dex_filenames;
    Split(class_path_string_, ':', &dex_filenames);
    RegisterAppInfo(dex_filenames,
        jit_options_->GetProfileSaverOptions().GetProfilePath());
}

return true;
}
```

Runtime 的 Init() 函数和 Start() 函数的内部都包含了很多重要的信息和操作，但是鉴于篇幅有限，不在此处介绍。在后续的部分中，会在相关内容中进行具体介绍。

Runtime 在 Start() 函数执行完之后，就被认为已经启动成功了，也就是说只要 Java 虚拟机启动成功，ART 的核心部分就启动成功了。Runtime::Start() 函数中涉及的内容比较多，会在后续的内容中对其中的各部分进行介绍。

5.3　ART 启动中的 JIT 编译器的创建

Dalvik 虚拟机中包含了 JIT 编译器，目前的 ART 中依然保留了 JIT 编译的功能。在 ART 启动阶段中，也会创建一个 JIT 编译器，本部分内容将对 JIT 编译器创建的流程进行分析。

在上文所提到的Runtime::Start()函数中包含了对JIT编译器的创建,这是为了保障在运行时需要进行JIT编译的时候,有对应的编译器可用。下面将对Runtime::Start()函数中所调用的JIT编译器的创建过程进行介绍。

在Runtime类的Start()函数中,有对CreateJit()函数的调用,该函数最终是为了创建JIT编译器。Runtime::CreateJit()函数的实现也位于art/Runtime/Runtime.cc中,它通过调用Jit::Create()函数新建Jit类,代码如下:

```
//第5章/Runtime.cc
void Runtime::CreateJit() {
  DCHECK(jit_ == nullptr);
  if (jit_code_cache_.get() == nullptr) {
    if (!IsSafeMode()) {
      LOG(WARNING) << "Missing code cache, cannot create JIT.";
    }
    return;
  }
  if (IsSafeMode()) {
    LOG(INFO) << "Not creating JIT because of SafeMode.";
    jit_code_cache_.reset();
    return;
  }

  jit::Jit* jit = jit::Jit::Create(jit_code_cache_.get(),
      jit_options_.get());
  DoAndMaybeSwitchInterpreter([=](){ jit_.reset(jit); });
  if (jit == nullptr) {
    LOG(WARNING) << "Failed to allocate JIT";
    //Release JIT code cache resources (several MB of memory)
    jit_code_cache_.reset();
  } else {
    jit->CreateThreadPool();
  }
}
```

其中,Jit::Create()函数的实现位于art/Runtime/jit/jit.cc中,是Jit类的成员函数。它通过对jit_load_()函数的执行创建了一个新的JitCompiler(jit_compiler_handle_),并且该JitCompiler的_compiler变量被设置成一个新建的OptimizingCompiler。之后,还新建了一个Jit类。Jit::Create()函数的代码如下:

```
//第5章/jit.cc
Jit* Jit::Create(JitCodeCache* code_cache, JitOptions* options) {
```

```cpp
  if (jit_load_ == nullptr) {
    LOG(WARNING) << "Not creating JIT: library not loaded";
    return nullptr;
  }
  jit_compiler_handle_ = (jit_load_)();
  if (jit_compiler_handle_ == nullptr) {
    LOG(WARNING) << "Not creating JIT: failed to allocate a compiler";
    return nullptr;
  }
  std::unique_ptr<Jit> jit(new Jit(code_cache, options));

  //If the code collector is enabled, check if that still holds
  //With 'perf', we want a 1-1 mapping between an address and a method
  //We aren't able to keep method pointers live during the
  //instrumentation method entry trampoline so we will just disable
  //jit-gc if we are doing that
  if (code_cache->GetGarbageCollectCode()) {
    code_cache->SetGarbageCollectCode(
        !jit_generate_debug_info_(jit_compiler_handle_) &&
        !Runtime::Current()->GetInstrumentation()
            ->AreExitStubsInstalled());
  }

  VLOG(jit) << "JIT created with initial_capacity="
      << PrettySize(options->GetCodeCacheInitialCapacity())
      << ", max_capacity="
      << PrettySize(options->GetCodeCacheMaxCapacity())
      << ", compile_threshold=" << options->GetCompileThreshold()
      << ", profile_saver_options=" << options->GetProfileSaverOptions();

  //Notify native debugger about the classes already loaded before the
  //creation of the jit
  jit->DumpTypeInfoForLoadedTypes(Runtime::Current()->GetClassLinker());
  return jit.release();
}
```

其中的 jit_load_() 函数是通过 Runtime::Start() 函数中调用的 Jit::LoadCompilerLibrary() 函数进行挂载的，其实际执行的函数位于 art/compiler/jit/jit_compiler.cc 文件中，这个函数的代码如下：

```cpp
//第5章/jit_compiler.cc
extern "C" void* jit_load() {
  VLOG(jit) << "Create jit compiler";
  auto* const jit_compiler = JitCompiler::Create();
  CHECK(jit_compiler != nullptr);
```

```
    VLOG(jit) << "Done creating jit compiler";
    return jit_compiler;
}
```

其中，Jitcompiler::Create()函数通过调用 JitCompiler 类的构造函数，来生成新的 Jitcompiler 对象。Jitcompiler::Create()函数也位于 art/compiler/jit/jit_compiler.cc 文件中，代码如下：

```
//第 5 章/ jit_compiler.cc
JitCompiler * JitCompiler::Create() {
    return new JitCompiler();
}
```

Jitcompiler 的构造函数也位于 art/compiler/jit/jit_compiler.cc 文件中，代码如下：

```
//第 5 章/ jit_compiler.cc
JitCompiler::JitCompiler() {
    compiler_options_.reset(new CompilerOptions());
    ParseCompilerOptions();
    compiler_.reset(
            Compiler:: Create ( * compiler _ options _, / * storage = */ nullptr, Compiler::
kOptimizing));
}
```

JitCompiler::JitCompiler()函数通过 Compiler::Create()函数最终为 std::unique_ptr <Compiler>类型的 compiler_重新赋值。Compiler::Create()函数的实现位于 art/compiler/compiler.cc 中，代码如下：

```
//第 5 章/ compiler.cc
Compiler * Compiler::Create(
    const CompilerOptions& compiler_options,
    CompiledMethodStorage * storage, Compiler::Kind kind) {
  //Check that oat version when Runtime was compiled matches the oat
  //version of the compiler.
  constexpr std::array< uint8_t, 4 > compiler_oat_version = OatHeader::kOatVersion;
  OatHeader::CheckOatVersion(compiler_oat_version);
  switch (kind) {
    case kQuick:
      //TODO: Remove Quick in options.
    case kOptimizing:
      return CreateOptimizingCompiler(compiler_options, storage);

    default:
      LOG(FATAL) << "UNREACHABLE";
```

```
        UNREACHABLE();
    }
}
```

因为 Compiler::Create() 接收了 Compiler::kOptimizing 参数,所以这里最终会调用 CreateOptimizingCompiler() 函数。CreateOptimizingCompiler() 函数的实现位于 art/compiler/optimizing/optimizing_compiler.cc 中,代码如下:

```
//第5章/ optimizing_compiler.cc
Compiler* CreateOptimizingCompiler(
    const CompilerOptions& compiler_options,
    CompiledMethodStorage* storage) {
  return new OptimizingCompiler(compiler_options, storage);
}
```

最终,返回的是一个 OptimizingCompiler 对象,也就是说这里的 JitCompiler 的 compiler_指针,指向一个新建的 OptimizingCompiler 对象。有关 OptimizingCompiler 的相关内容,在之前的内容进行过介绍。

这里涉及的 JitCompiler 最终还是一个 OptimizingCompiler,它们都是 art/compiler 部分的内容,所以在 Jit::LoadCompilerLibrary() 函数中加载的也是 libartd-compiler.so。

总结本节的内容,Android Runtime 在 Jit 运行的时候,还是调用了 art/compiler 内的 OptimizingCompiler,这和 dex2oat 调用的 OptimizingCompiler 是一回事,因此,也可以将 OptimizingCompiler 理解为调用的是 ART 本身最主要的 Compiler 体系。

5.4 ART 启动中的 Thread 处理

Thread 是 ART 中的重要组成部分。在 ART 的启动阶段,也有与 Thread 类相关的重要操作。ART 启动过程中有两个顺序执行的重要函数,即 Runtime::Init() 函数和 Runtime::Start() 函数,这两个函数的实现都位于 art/Runtime/Runtime.cc 中,它们中包含了在 ART 启动阶段有关 Thread 类的主要操作,本部分将对其进行介绍。

在 Runtime::Init() 函数中,有关 Thread 类的重要操作主要有 Thread::Startup() 函数和 Thread::Attach() 函数,代码如下:

```
//第5章/Runtime.cc
bool Runtime::Init(RuntimeArgumentMap&& Runtime_options_in) {
    …
    Thread::Startup();

    //ClassLinker needs an attached thread, but we can't fully attach a
```

```
    //thread without creating objects. We can't supply a thread group yet
    //it will be fixed later. Since we are the main thread, we do not get
    //a java peer
    Thread* self = Thread::Attach("main", false, nullptr, false);
    CHECK_EQ(self->GetThreadId(), ThreadList::kMainThreadId);
    CHECK(self != nullptr);

    self->SetIsRuntimeThread(IsAotCompiler());

    //Set us to runnable so tools using a Runtime can allocate and GC by
    //default
    self->TransitionFromSuspendedToRunnable();
    …
    return true;
}
```

在 Runtime::Start()函数中,有关 Thread 类的重要操作有 InitThreadGroups()函数、Thread::FinishStartup()函数和 StartDaemonThreads()函数,具体的代码如下:

```
//第5章/Runtime.cc
bool Runtime::Start() {
    VLOG(startup) << "Runtime::Start entering";
…
    //Restore main thread state to kNative as expected by native code
    Thread* self = Thread::Current();
…

    //Initialize well known thread group values that may be accessed
    //threads while attaching
    InitThreadGroups(self);

    Thread::FinishStartup();
…
    StartDaemonThreads();
…
    return true;
}
```

StartDaemonThreads()函数的实现位于 art/Runtime/Runtime.cc 中,它是 Runtime 的成员函数。Runtime::StartDaemonThreads()函数中的核心代码调用了 JNIEnv 的 CallStaticVoidMethod()函数,代码如下:

```
//第5章/Runtime.cc
void Runtime::StartDaemonThreads() {
  ScopedTrace trace(__FUNCTION__);
  VLOG(startup) << "Runtime::StartDaemonThreads entering";

  Thread* self = Thread::Current();

  //Must be in the kNative state for calling native methods
  CHECK_EQ(self->GetState(), kNative);

  JNIEnv* env = self->GetJniEnv();
  env->CallStaticVoidMethod(WellKnownClasses::java_lang_Daemons,
                            WellKnownClasses::java_lang_Daemons_start);
  if (env->ExceptionCheck()) {
    env->ExceptionDescribe();
    LOG(FATAL) << "Error starting java.lang.Daemons";
  }

  VLOG(startup) << "Runtime::StartDaemonThreads exiting";
}
```

在 StartDaemonThreads() 函数中，除了核心调用 JNIEnv 的 CallStaticVoidMethod() 函数之外，剩余的代码主要是 trace、VLOG 和异常处理。这里涉及的 JNIEnv 是一个很重要的类型，它的定义位于 libnativehelper/include_jni/jni.h 中，具体的代码如下：

```
//第5章/jni.h
#if defined(__cplusplus)
typedef _JNIEnv JNIEnv;
typedef _JavaVM JavaVM;
#else
typedef const struct JNINativeInterface* JNIEnv;
typedef const struct JNIInvokeInterface* JavaVM;
#endif
```

从上述代码可以看出，JNIEnv 可能有两种情况，一种是 _JNIEnv，另一种是 JNINativeInterface*，这两者的定义也都位于 jni.h 中。_JNIEnv 是一个结构体，它可以被视为一个包装类，或者一个转换器，其内部的函数调用都被转换到 JNINativeInterface* 指向的对象的函数调用，StartDaemonThreads() 函数中调用的 JNIEnv 的 CallStaticVoidMethod() 函数也不例外。_JNIEnv 的代码如下：

```
//第5章/jni.h
/*
 * C++ object wrapper.
```

```
 *
 * This is usually overlaid on a C struct whose first element is a
 * JNINativeInterface*. We rely somewhat on compiler behavior.
 */
struct _JNIEnv {
    /* do not rename this; it does not seem to be entirely opaque */
    const struct JNINativeInterface* functions;

#if defined(__cplusplus)

    jint GetVersion()
    { return functions->GetVersion(this); }

    jclass DefineClass(const char *name, jobject loader,
        const jByte* buf, jsize bufLen)
    { return functions->DefineClass(this, name, loader, buf, bufLen); }

    …

    void CallStaticVoidMethod(jclass clazz, jmethodID methodID, ...)
    {
        va_list args;
        va_start(args, methodID);
        functions->CallStaticVoidMethodV(this, clazz, methodID, args);
        va_end(args);
    }
    void CallStaticVoidMethodV(jclass clazz, jmethodID methodID,
        va_list args)
    { functions->CallStaticVoidMethodV(this, clazz, methodID, args); }
    void CallStaticVoidMethodA(jclass clazz, jmethodID methodID,
        const jvalue* args)
    { functions->CallStaticVoidMethodA(this, clazz, methodID, args); }

    …
}
```

JNIEnv 的 CallStaticVoidMethod()函数的调用,在上述代码中被转换为 functions->CallStaticVoidMethod()函数的调用,而 functions 正是 JNINativeInterface *。至此,可以得知,无论 JNIEnv 是_JNIEnv,还是 JNINativeInterface *,对于 CallStaticVoidMethod()函数的调用都会被转化为对 JNINativeInterface * 类型所指向的 CallStaticVoidMethod()函数。JNINativeInterface 其实也是一个结构体,它是一个接口函数的指针列表,它有一系列的函数指针,其中就包括指向 CallStaticVoidMethod()函数的函数指针,代码如下:

```
//第5章/jni.h
/*
 * Table of interface function pointers.
```

```
 */
struct JNINativeInterface {
    void *      reserved0;
    void *      reserved1;
    void *      reserved2;
    void *      reserved3;

    jint        (*GetVersion)(JNIEnv *);
...
    void        (*CallStaticVoidMethod)(JNIEnv *, jclass, jmethodID, ...);
    void        (*CallStaticVoidMethodV)(JNIEnv *, jclass, jmethodID,
        va_list);
    void        (*CallStaticVoidMethodA)(JNIEnv *, jclass, jmethodID, const jvalue *);

    jfieldID    (*GetStaticFieldID)(JNIEnv *, jclass, const char *,
        const char *);
...
}
```

对于JNINativeInterface这个结构的实例化,发生在art/Runtime/jni/jni_internal.cc中,定义了一个gJniNativeInterface,其中对于JNINativeInterface结构体中的所有函数指针都进行了挂载,包括CallStaticVoidMethod()函数,代码如下:

```
//第5章/jni_internal.cc
const JNINativeInterface gJniNativeInterface = {
  nullptr, //reserved0
  nullptr, //reserved1
  nullptr, //reserved2
  nullptr, //reserved3
  JNI::GetVersion,
  JNI::DefineClass,

  ...

  JNI::CallStaticVoidMethod,
  JNI::CallStaticVoidMethodV,
  JNI::CallStaticVoidMethodA,

  ...
};
```

所以,对于JNIEnv的CallStaticVoidMethod()函数的调用,最终会被转换为对JNI::CallStaticVoidMethod()函数的调用。JNI::CallStaticVoidMethod()函数的实现位于art/Runtime/jni/jni_internal.cc中,代码如下:

```
//第 5 章/jni_internal.cc
    static void CallStaticVoidMethod(JNIEnv* env, jclass,
        jmethodID mid, ...) {
    va_list ap;
    va_start(ap, mid);
    ScopedVAArgs free_args_later(&ap);
    CHECK_NON_NULL_ARGUMENT_RETURN_VOID(mid);
    ScopedObjectAccess soa(env);
    InvokeWithVarArgs(soa, nullptr, mid, ap);
}
```

其中，env 的值就是最早在 StartDaemonThreads()函数中的 env，jclass 类型参数的值为 WellKnownClasses::java_lang_Daemons，mid 的值为 WellKnownClasses::java_lang_Daemons_start。JNI::CallStaticVoidMethod()函数的 3 个值其实都是在 StartDaemonThreads()函数中被确定了，但是仅仅使用了 env 和 mid，需要特别记住 mid 的值，这是后续处理的关键信息。JNI::CallStaticVoidMethod()函数中所调用的 InvokeWithVarArgs()函数的实现位于 art/Runtime/reflection.cc 中，代码如下：

```
//第 5 章/reflection.cc
JValue InvokeWithVarArgs(const ScopedObjectAccessAlreadyRunnable& soa,
    jobject obj, jmethodID mid, va_list args)
    REQUIRES_SHARED(Locks::mutator_lock_) {
//We want to make sure that the stack is not within a small distance
//from the protected region in case we are calling into a leaf
//function whose stack check has been elided
if (UNLIKELY(__builtin_frame_address(0) < soa.Self()->GetStackEnd())) {
    ThrowStackOverflowError(soa.Self());
    return JValue();
}

ArtMethod* method = jni::DecodeArtMethod(mid);
bool is_string_init = method->GetDeclaringClass()->IsStringClass() && method->IsConstructor();
if (is_string_init) {
    //Replace calls to String.<init> with equivalent StringFactory call
    method = WellKnownClasses::StringInitToStringFactory(method);
}
ObjPtr<mirror::Object> receiver = method->IsStatic() ? nullptr : soa.Decode<mirror::Object>(obj);
uint32_t shorty_len = 0;
const char* shorty =
    method->GetInterfaceMethodIfProxy(kRuntimePointerSize)
        ->GetShorty(&shorty_len);
JValue result;
```

```
    ArgArray arg_array(shorty, shorty_len);
    arg_array.BuildArgArrayFromVarArgs(soa, receiver, args);
    InvokeWithArgArray(soa, method, &arg_array, &result, shorty);
    if (is_string_init) {
      //For string init, remap original receiver to StringFactory result.
      UpdateReference(soa.Self(), obj, result.GetL());
    }
    return result;
}
```

其中,比较关键的是 DecodeArtMethod()函数和 InvokeWithArgArray()函数。下面对这两个函数依次进行介绍。DecodeArtMethod()函数根据 mid,也就是根据 WellKnownClasses::java_lang_Daemons_start 构建了一个 ArtMethod * 指针。它的代码位于 art/Runtime/jni/jni_internal.h 中,代码如下:

```
//第 5 章/jni_internal.h
ALWAYS_INLINE
static inline ArtMethod* DecodeArtMethod(jmethodID method_id) {
  return reinterpret_cast<ArtMethod*>(method_id);
}
```

这里所提到的 WellKnownClasses::java_lang_Daemons_start 的定义位于 art/Runtime/well_know_classes.cc 中,所有的 WellKnownClasses 的 jmethodID 都在这里,此外 jclass 和 jfieldID 也都在这里。java_lang_Daemons_start 是在 WellKnownClasses::Init() 函数中获取的值,代码如下:

```
//第 5 章/well_know_classes.cc
void WellKnownClasses::Init(JNIEnv* env) {
  hiddenapi::ScopedHiddenApiEnforcementPolicySetting hiddenapi_exemption(
      hiddenapi::EnforcementPolicy::kDisabled);
  ...
  java_lang_Daemons = CacheClass(env, "java/lang/Daemons");
  ...
  java_lang_Daemons_start = CacheMethod(env, java_lang_Daemons, true, "start", "()V");
  java_lang_Daemons_stop = CacheMethod(env, java_lang_Daemons, true, "stop", "()V");
  ...
}
```

其中 CacheMethod()函数通过输入的几个参数缓存了 java/lang/Daemons 类的 start()方法,并且传递给 WellKnownClasses::java_lang_Daemons_start()这种方法的 jmethodID,但还是 static 类型的,代码如下:

```cpp
//第5章/well_know_classes.cc
static jmethodID CacheMethod(JNIEnv* env, jclass c, bool is_static,
    const char* name, const char* signature) {
  jmethodID mid = is_static ? env->GetStaticMethodID(c, name, ignature) :
      env->GetMethodID(c, name, signature);
  if (mid == nullptr) {
    ScopedObjectAccess soa(env);
    if (soa.Self()->IsExceptionPending()) {
      LOG(FATAL_WITHOUT_ABORT) << soa.Self()->GetException()->Dump();
    }
    std::ostringstream os;
    WellKnownClasses::ToClass(c)->DumpClass(os, mirror::Class::kDumpClassFullDetail);
    LOG(FATAL) << "Couldn't find method \"" << name
               << "\" with signature \"" << signature << "\": "
               << os.str();
  }
  return mid;
}
```

接下来分析 InvokeWithArgArray() 函数，InvokeWithArgArray() 函数也位于 art/Runtime/reflection.cc 中，它通过调用 ArtMethod 的 Invoke() 函数，进行下一步的操作，代码如下：

```cpp
//第5章/reflection.cc
void InvokeWithArgArray(const ScopedObjectAccessAlreadyRunnable& soa,
    ArtMethod* method, ArgArray* arg_array, JValue* result,
    const char* shorty) REQUIRES_SHARED(Locks::mutator_lock_) {
  uint32_t* args = arg_array->GetArray();
  if (UNLIKELY(soa.Env()->IsCheckJniEnabled())) {
    CheckMethodArguments(soa.Vm(), method->GetInterfaceMethodIfProxy(kRuntimePointerSize), args);
  }
  method->Invoke(soa.Self(), args, arg_array->GetNumBytes(), result, shorty);
}
```

ArtMethod::Invoke() 函数的实现位于 art/Runtime/art_method.cc 中，其中最重要的是要区分是解释执行还是调用 invoke stub，代码如下：

```cpp
//第5章/art_method.cc
void ArtMethod::Invoke(Thread* self, uint32_t* args, uint32_t args_size,
    JValue* result, const char* shorty) {
  if (UNLIKELY(__builtin_frame_address(0) < self->GetStackEnd())) {
    ThrowStackOverflowError(self);
    return;
```

```cpp
    }

    if (kIsDebugBuild) {
      self->AssertThreadSuspensionIsAllowable();
      CHECK_EQ(kRunnable, self->GetState());
      CHECK_STREQ(GetInterfaceMethodIfProxy(kRuntimePointerSize)
          ->GetShorty(), shorty);
    }

    //Push a transition back into managed code onto the linked list in
    //thread
    ManagedStack fragment;
    self->PushManagedStackFragment(&fragment);

    Runtime* Runtime = Runtime::Current();
    //Call the invoke stub, passing everything as arguments
    //If the Runtime is not yet started or it is required by the debugger
    //then perform the Invocation by the interpreter, explicitly forcing
    //interpretation over JIT to prevent cycling around the various
    //JIT/Interpreter methods that handle method invocation
    if (UNLIKELY(!Runtime->IsStarted() || (self->IsForceInterpreter()
        && !IsNative() && !IsProxyMethod() && IsInvokable()) ||
        Dbg::IsForcedInterpreterNeededForCalling(self, this))) {
      if (IsStatic()) {
        art::interpreter::EnterInterpreterFromInvoke(self, this, nullptr,
            args, result, /* stay_in_interpreter = */ true);
      } else {
        mirror::Object* receiver = reinterpret_cast<StackReference<mirror::Object>*>
            (&args[0])->AsMirrorPtr();
        art::interpreter::EnterInterpreterFromInvoke(self, this, receiver,
            args + 1, result, /* stay_in_interpreter = */ true);
      }
    } else {
      DCHECK_EQ(Runtime->GetClassLinker()->GetImagePointerSize(), kRuntimePointerSize);

      constexpr bool kLogInvocationStartAndReturn = false;
      bool have_quick_code = GetEntryPointFromQuickCompiledCode() != nullptr;
      if (LIKELY(have_quick_code)) {
        if (kLogInvocationStartAndReturn) {
          LOG(INFO) << StringPrintf("Invoking '%s' quick code = %p static = %d", PrettyMethod().c_str(), GetEntryPointFromQuickCompiledCode(), static_cast<int>(IsStatic() ? 1 : 0));
        }

        //Ensure that we won't be accidentally calling quick compiled code
        //when -Xint
        if (kIsDebugBuild && Runtime->GetInstrumentation()->IsForcedInterpretOnly()) {
```

```cpp
      CHECK(!Runtime->UseJitCompilation());
      const void* oat_quick_code = (IsNative() || !IsInvokable() ||
          IsProxyMethod() || IsObsolete()) ? nullptr
          : GetOatMethodQuickCode(Runtime->GetClassLinker()
              ->GetImagePointerSize());
      CHECK(oat_quick_code == nullptr || oat_quick_code != GetEntryPointFromQuickCompiledCode())
          << "Don't call compiled code when -Xint " << PrettyMethod();
    }

    if (!IsStatic()) {
      (*art_quick_invoke_stub)(this, args, args_size, self, result, shorty);
    } else {
      (*art_quick_invoke_static_stub)(this, args, args_size, self, result, shorty);
    }
    if (UNLIKELY(self->GetException() == Thread::GetDeoptimizationException())) {
      //Unusual case where we were running generated code and an
      //exception was thrown to force the activations to be removed
      //from the stack. Continue execution in the interpreter
      self->DeoptimizeWithDeoptimizationException(result);
    }
    if (kLogInvocationStartAndReturn) {
      LOG(INFO) << StringPrintf("Returned '%s' quick code =%p", PrettyMethod().c_str(),
GetEntryPointFromQuickCompiledCode());
    }
  } else {
    LOG(INFO) << "Not invoking '" << PrettyMethod() << "' code=null";
    if (result != nullptr) {
      result->SetJ(0);
    }
  }
}

//Pop transition
self->PopManagedStackFragment(fragment);
}
```

ArtMethod::Invoke()函数是一个非常重要的入口函数,并不仅在这个场景下使用,它作为 ArtMethod 类的重要函数,还会在其他场景使用。同时,根据条件判断,Invoke()函数中最终有可能出现 4 种走向。第一层次的判断是 Runtime,如果没有 start,或者需要 debug 等条件下,这种情况下则进行解释执行(包含 JIT);在进入解释执行选择之后,还会根据 ArtMethod 是否为 static 最终为 EnterInterpreterFromInvoke 构建参数,然后进行调用。否则,如果不进入解释执行模式,则通过 invoke stub,通过判断是否为 static,将所有的参数传递给 art_quick_invoke_stub()函数或者 art_quick_invoke_static_stub()函数的指针所指向的函数。守望线程在非 debug 模式之下,它执行到这里的时候,Runtime 已经准备好了,并且它也是 native(WellKnownClasses 里存放的都是 native methods)且是 static 的,这些内容都可以通过回溯之前的代码实现,那么在这种情况之下,在这里最终会执行 art_quick_invoke_static_stub()函数的指针所指向的函数。

art_quick_invoke_static_stub()函数和 art_quick_invoke_stub()函数的指针所指向的函数,最终在实现上位于 Runtime/arch/XXX/quick_entrypoints_XXX.S 中。以 x86 的 64 位架构为例,则位于 Runtime/arch/x86/quick_entrypoints_x86.S 中。通过这里的 art_quick_invoke_static_stub()函数,接收在 ArtMethod::Invoke()函数中传递进去的所有参数。

从上文可知,这时候传递进去的 ArtMethod 实际上是 java_lang_Daemons 的 start()函数,而 java_lang_Daemons 这个 jclass 则对应的是 java/lang/Daemons。java/lang/Daemons 位于 libcore/libart/src/main/java/java/lang/Daemons.java 中,类 Daemons 中还含有一个抽象类 Daemon 及其 4 个子类 HeapTaskDaemon、ReferenceQueueDaemon、FinalizerDaemon 和 FinalizerWatchdogDaemon。Daemons 的 start()函数则分别调用了 Daemon 这 4 个子类的 start()函数,代码如下:

```
//第 5 章/Daemons.java
    public static void start() {
        for (Daemon daemon : DAEMONS) {
            daemon.start();
        }
    }
```

其中的 DAEMONS 就是 Daemon 这 4 个子类所组成的数组,代码如下:

```
//第 5 章/Daemons.java
    private static final Daemon[] DAEMONS = new Daemon[] {
            HeapTaskDaemon.INSTANCE,
            ReferenceQueueDaemon.INSTANCE,
            FinalizerDaemon.INSTANCE,
            FinalizerWatchdogDaemon.INSTANCE,
    };
```

Daemon 这 4 个子类并没有 start()函数,但是 Daemon 有 start()函数,代码如下:

```
//第 5 章/Daemons.java
        public synchronized void start() {
            startInternal();
        }
```

Daemon 这 4 个子类同样没有 startInternal()函数,所以将执行 Daemon 的 startInternal()函数,新建新的线程,代码如下:

```
//第5章/Daemons.java
    public void startInternal() {
        if (thread != null) {
            throw new IllegalStateException("already running");
        }
        thread = new Thread(ThreadGroup.systemThreadGroup, this, name);
        thread.setDaemon(true);
        thread.setSystemDaemon(true);
        thread.start();
    }
```

这部分内容其实是一个典型的底层代码(C/C++)通过反射机制调用上层代码(Java)的例子,这是 JNI 的典型情况之一。JNI 通常分为上层代码调用底层代码和底层代码调用上层代码这两种情况,所以 JNI 是一个双向的调用接口,日常常见的是上层代码调用底层代码,这里展现的是另外一种情况。

至此,已经梳理了启动过程中的线程处理,特别是其中的守护线程启动的流程。这部分内容对于理解 ART 线程的启动十分重要,其中有些内容介绍得比较简单,读者可以根据已经介绍的部分内容所对应的代码进行展开。

5.5 ART 启动中的运行时本地方法初始化

在上文介绍守护线程部分的时候涉及了有关 java/lang/Daemons 类的缓存,并且最后还执行了它的 start() 方法。考虑到介绍的一致性,上文并没有对这类与运行时相关的 JNI 本地方法初始化进行介绍,本部分将对 ART 启动中的与运行时相关的本地方法初始化做一个介绍。

在 ART 的启动过程中,与运行时相关的本地方法的初始化发生在守护线程的启动之前,这点可以在 Runtime::Start() 函数(art/Runtime/Runtime.cc)中看到,本地方法的初始化是在 InitNativeMethods() 函数中执行的,它位于守护线程的 StartDaemonThreads() 函数之前,代码如下:

```
//第5章/ Runtime.cc
bool Runtime::Start() {
  VLOG(startup) << "Runtime::Start entering";

  CHECK(!no_sig_chain_) << "A started Runtime should have sig chain enabled";
  …
  //InitNativeMethods needs to be after started_ so that the classes
  //it touches will have methods linked to the oat file if necessary
  {
    ScopedTrace trace2("InitNativeMethods");
```

```
    InitNativeMethods();
  }
  …
  StartDaemonThreads();

  //Make sure the environment is still clean (no lingering local refs
  //from starting daemon threads)
  {
    ScopedObjectAccess soa(self);
    self->GetJniEnv()->AssertLocalsEmpty();
  }
  …

  return true;
}
```

InitNativeMethods()函数的实现也在 RunTime.cc 中,其中所执行的内容可以分为注册运行时的本地方法、初始化 well known 类(被 JNI 使用的类)、加载 libjavacore 和 libopenJDK、初始化 well known 类(调用本地方法的类)、初始化本地代码通过 JNI 调用 Java 代码的检查。这里注释涉及的 native method 指的是 Java 代码通过 JNI 调用的 C/C++方法;涉及的 managed code 指的是用 Java 实现的代码。InitNativeMethods()函数的代码如下:

```
//第 5 章/ Runtime.cc
void Runtime::InitNativeMethods() {
  VLOG(startup) << "Runtime::InitNativeMethods entering";
  Thread * self = Thread::Current();
  JNIEnv * env = self->GetJniEnv();

  //Must be in the kNative state for calling native methods (JNI_OnLoad
  //code)
  CHECK_EQ(self->GetState(), kNative);

  //Set up the native methods provided by the Runtime itself
  RegisterRuntimeNativeMethods(env);

  //Initialize classes used in JNI. The initialization requires Runtime
  //native methods to be loaded first
  WellKnownClasses::Init(env);

  //Then set up libjavacore / libopenJDK, which are just a regular JNI
  //libraries with a regular JNI_OnLoad. Most JNI libraries can just use
  //System.loadLibrary, but libcore can't because it's the library that
  //implements System.loadLibrary
  {
    std::string error_msg;
```

```
    if (!java_vm_ -> LoadNativeLibrary(env, "libjavacore.so", nullptr,
        WellKnownClasses::java_lang_Object, &error_msg)) {
      LOG(FATAL) << "LoadNativeLibrary failed for \"libjavacore.so\": "
        << error_msg;
    }
  }
  {
    constexpr const char* kOpenJDKLibrary = kIsDebugBuild
        ? "libopenJDKd.so" : "libopenJDK.so";
    std::string error_msg;
    if (!java_vm_ -> LoadNativeLibrary(env, kOpenJDKLibrary, nullptr,
        WellKnownClasses::java_lang_Object, &error_msg)) {
      LOG(FATAL) << "LoadNativeLibrary failed for \""
                 << kOpenJDKLibrary << "\": " << error_msg;
    }
  }

  //Initialize well known classes that may invoke Runtime native
  //methods
  WellKnownClasses::LateInit(env);

  //Having loaded native libraries for Managed Core library, enable
  //field and method resolution checks via JNI from native code
  JniInitializeNativeCallerCheck();

  VLOG(startup) << "Runtime::InitNativeMethods exiting";
}
```

其中，RegisterRuntimeNativeMethods()函数负责注册运行时的本地方法，其函数内部是一系列的注册本地方法的操作，主要根据类别进行注册，代码如下：

```
//第 5 章/ Runtime.cc
void Runtime::RegisterRuntimeNativeMethods(JNIEnv* env) {
  register_dalvik_system_DexFile(env);
  register_dalvik_system_VMDebug(env);
  register_dalvik_system_VMRuntime(env);
  register_dalvik_system_VMStack(env);
  register_dalvik_system_ZygoteHooks(env);
  register_java_lang_Class(env);
  register_java_lang_Object(env);
  register_java_lang_invoke_MethodHandleImpl(env);
  register_java_lang_ref_FinalizerReference(env);
  register_java_lang_reflect_Array(env);
  register_java_lang_reflect_Constructor(env);
  register_java_lang_reflect_Executable(env);
```

```
    register_java_lang_reflect_Field(env);
    register_java_lang_reflect_Method(env);
    register_java_lang_reflect_Parameter(env);
    register_java_lang_reflect_Proxy(env);
    register_java_lang_ref_Reference(env);
    register_java_lang_String(env);
    register_java_lang_StringFactory(env);
    register_java_lang_System(env);
    register_java_lang_Thread(env);
    register_java_lang_Throwable(env);
    register_java_lang_VMClassLoader(env);
    register_java_util_concurrent_atomic_AtomicLong(env);
    register_libcore_util_CharsetUtils(env);
    register_org_apache_harmony_dalvik_ddmc_DdmServer(env);
    register_org_apache_harmony_dalvik_ddmc_DdmVmInternal(env);
    register_sun_misc_Unsafe(env);
}
```

这里面虽然类别很多,但是其运行模式都是相同的,所以才用了相同的运行模式对不同类别的本地方法进行了注册,这样便可搞清楚其中一个类别的本地方法的运行模式,也就搞清楚了全部。这里取第 1 个 register_dalvik_system_DexFile() 函数进行分析。register_dalvik_system_DexFile() 函数的实现位于 art/Runtime/native/dalvik_system_DexFile.cc 中,代码如下:

```
//第 5 章/ dalvik_system_DexFile.cc
void register_dalvik_system_DexFile(JNIEnv* env) {
    REGISTER_NATIVE_METHODS("dalvik/system/DexFile");
}
```

这里调用了 REGISTER_NATIVE_METHODS 宏,它的定义位于 art/Runtime/native/native_util.h 中,代码如下:

```
//第 5 章/native_util.h
#define REGISTER_NATIVE_METHODS(jni_class_name) \
    RegisterNativeMethodsInternal(env, (jni_class_name), gMethods, \
    arraysize(gMethods))
}
```

所调用的 RegisterNativeMethodsInternal() 函数也位于 native_util.h 中,代码如下:

```
//第 5 章/native_util.h
ALWAYS_INLINE inline void RegisterNativeMethodsInternal(JNIEnv* env,
    const char* jni_class_name, const JNINativeMethod* methods,
    jint method_count) {
    ScopedLocalRef<jclass> c(env, env->FindClass(jni_class_name));
```

```
    if (c.get() == nullptr) {
        LOG(FATAL) << "Couldn't find class: " << jni_class_name;
    }
    jint jni_result = env->RegisterNatives(c.get(), methods, method_count);
    CHECK_EQ(JNI_OK, jni_result);
}
```

register_dalvik_system_DexFile()函数通过 REGISTER_NATIVE_METHODS 宏最终调用了 RegisterNativeMethodsInternal()函数，并且为其传递了 env、jni_class_name、gMethods 和 gMethods 中方法数量。这里的 env 是从上层一直传递进来的，jni_class_name 在这里指的是具体的 "dalvik/system/DexFile"，gMethods 指的是 art/Runtime/native/dalvik_system_DexFile.cc 中所定义的 gMethods。RegisterNativeMethodsInternal()函数调用了 JNIEnv 的 RegisterNatives()函数。前文介绍过，JNIEnv 的函数在实际操作中最终会转化为 JNI 的同名函数，所以最终调用了 JNI 的 RegisterNatives()函数，它位于 art/Runtime/jni/jni_internal.cc 中，主要通过 JNINativeMethod 中所包含的名字（name）和签名（signature）去构建 ArtMethod，并向 ArtMethod 注册函数指针 fnPtr，代码如下：

```
//第5章/jni_internal.cc
static jint RegisterNatives(JNIEnv* env, jclass java_class,
    const JNINativeMethod* methods, jint method_count) {
    if (UNLIKELY(method_count < 0)) {
        JavaVmExtFromEnv(env)->JniAbortF("RegisterNatives", "negative method count: % d", method_count);
        return JNI_ERR; //Not reached except in unit tests
    }
    CHECK_NON_NULL_ARGUMENT_FN_NAME("RegisterNatives", java_class, JNI_ERR);
    ScopedObjectAccess soa(env);
    StackHandleScope<1> hs(soa.Self());
    Handle<mirror::Class> c = hs.NewHandle(
        soa.Decode<mirror::Class>(java_class));
    if (UNLIKELY(method_count == 0)) {
        LOG(WARNING) << "JNI RegisterNativeMethods: attempt to register 0 native methods for " <<
c->PrettyDescriptor();
        return JNI_OK;
    }
    CHECK_NON_NULL_ARGUMENT_FN_NAME("RegisterNatives", methods, JNI_ERR);
    for (jint i = 0; i < method_count; ++i) {
        const char* name = methods[i].name;
        const char* sig = methods[i].signature;
        const void* fnPtr = methods[i].fnPtr;
        if (UNLIKELY(name == nullptr)) {
            ReportInvalidJNINativeMethod(soa, c.Get(), "method name", i);
            return JNI_ERR;
```

```cpp
    } else if (UNLIKELY(sig == nullptr)) {
      ReportInvalidJNINativeMethod(soa, c.Get(), "method signature", i);
      return JNI_ERR;
    } else if (UNLIKELY(fnPtr == nullptr)) {
      ReportInvalidJNINativeMethod(soa, c.Get(), "native function", i);
      return JNI_ERR;
    }
    bool is_fast = false;
    //Notes about fast JNI calls:
    //
    //On a normal JNI call, the calling thread usually transitions
    //from the kRunnable state to the kNative state. But if the
    //called native function needs to access any Java object, it
    //will have to transition back to the kRunnable state
    //
    //There is a cost to this double transition. For a JNI call
    //that should be quick, this cost may dominate the call cost
    //
    //On a fast JNI call, the calling thread avoids this double
    //transition by not transitioning from kRunnable to kNative and
    //stays in the kRunnable state
    //
    //There are risks to using a fast JNI call because it can delay
    //a response to a thread suspension request which is typically
    //used for a GC root scanning, etc. If a fast JNI call takes a
    //long time, it could cause longer thread suspension latency
    //and GC pauses
    //
    //Thus, fast JNI should be used with care. It should be used
    //for a JNI call that takes a short amount of time (eg. no
    //long-running loop) and does not block (eg. no locks, I/O,
    //etc.)
    //
    //A '!' prefix in the signature in the JNINativeMethod
    //indicates that it's a fast JNI call and the Runtime omits the
    //thread state transition from kRunnable to kNative at the
    //entry
    if (*sig == '!') {
      is_fast = true;
      ++sig;
    }

    //Note: the right order is to try to find the method locally
    //first, either as a direct or a virtual method. Then move to
    //the parent
    ArtMethod* m = nullptr;
    bool warn_on_going_to_parent =
```

```cpp
        down_cast<JNIEnvExt*>(env)->GetVm()->IsCheckJniEnabled();
  for (ObjPtr<mirror::Class> current_class = c.Get();
       current_class != nullptr;
       current_class = current_class->GetSuperClass()) {
    //Search first only comparing methods which are native.
    m = FindMethod<true>(current_class, name, sig);
    if (m != nullptr) {
      break;
    }

    //Search again comparing to all methods, to find non-native
    //methods that match
    m = FindMethod<false>(current_class, name, sig);
    if (m != nullptr) {
      break;
    }

    if (warn_on_going_to_parent) {
      LOG(WARNING) << "CheckJNI: method to register \"" << name
                   << "\" not in the given class. "
                   << "This is slow, consider changing your RegisterNatives calls.";
      warn_on_going_to_parent = false;
    }
  }

  if (m == nullptr) {
    c->DumpClass(LOG_STREAM(ERROR),
        mirror::Class::kDumpClassFullDetail);
    LOG(ERROR)
        << "Failed to register native method "
        << c->PrettyDescriptor() << "." << name << sig << " in "
        << c->GetDexCache()->GetLocation()->ToModifiedUtf8();
    ThrowNoSuchMethodError(soa, c.Get(), name, sig,
                           "static or non-static");
    return JNI_ERR;
  } else if (!m->IsNative()) {
    LOG(ERROR)
        << "Failed to register non-native method "
        << c->PrettyDescriptor() << "." << name << sig
        << " as native";
    ThrowNoSuchMethodError(soa, c.Get(), name, sig, "native");
    return JNI_ERR;
  }

  VLOG(jni) << "[Registering JNI native method "
            << m->PrettyMethod() << "]";
```

```
    if (UNLIKELY(is_fast)) {
      //There are a few reasons to switch
      //1) We don't support !bang JNI anymore, it will turn to a hard
      //error later
      //2) @FastNative is actually faster. At least 1.5x faster
      //than !bang JNI
      //and switching is super easy, remove ! in C code, add annotation
      //in .java code
      //3) Good chance of hitting DCHECK failures in
      //ScopedFastNativeObjectAccess since that checks for presence of
      //@FastNative and not for ! in the descriptor
      LOG(WARNING) << "!bang JNI is deprecated.
          Switch to @FastNative for " << m->PrettyMethod();
      is_fast = false;
      //TODO: make this a hard register error in the future
    }

    const void* final_function_ptr = m->RegisterNative(fnPtr);
    UNUSED(final_function_ptr);
  }
  return JNI_OK;
}
```

这里用到了从 JNINativeMethod * methods 中获取 name、signature 和 fnPtr。JNINativeMethod 的定义位于 libnativehelper/include_jni/jni.h 中,代码如下:

```
//第5章/jni.h
typedef struct {
    const char* name;
    const char* signature;
    void*       fnPtr;
} JNINativeMethod;
```

这里的代码很清晰,JNINativeMethod 中保存了这 3 个指针。在这里,传递给 JNINativeMethod * methods 的实际内容是 art/Runtime/native/dalvik_system_DexFile.cc 中的 gMethods,gMethods 的代码如下:

```
//第5章/dalvik_system_DexFile.cc
static JNINativeMethod gMethods[] = {
  NATIVE_METHOD(DexFile, closeDexFile, "(Ljava/lang/Object;)Z"),
  NATIVE_METHOD(DexFile, defineClassNative,
                "(Ljava/lang/String;"
                "Ljava/lang/ClassLoader;"
                "Ljava/lang/Object;"
                "Ldalvik/system/DexFile;"
```

```
                    ")Ljava/lang/Class;"),
NATIVE_METHOD(DexFile, getClassNameList,
              "(Ljava/lang/Object;)[Ljava/lang/String;"),
NATIVE_METHOD(DexFile, isDexOptNeeded, "(Ljava/lang/String;)Z"),
NATIVE_METHOD(DexFile, getDexOptNeeded,
              "(Ljava/lang/String;Ljava/lang/String;"
              "Ljava/lang/String;Ljava/lang/String;ZZ)I"),
NATIVE_METHOD(DexFile, openDexFileNative,
              "(Ljava/lang/String;"
              "Ljava/lang/String;"
              "I"
              "Ljava/lang/ClassLoader;"
              "[Ldalvik/system/DexPathList$Element;"
              ")Ljava/lang/Object;"),
NATIVE_METHOD(DexFile, openInMemoryDexFilesNative,
              "([Ljava/nio/ByteBuffer;"
              "[[B"
              "[I"
              "[I"
              "Ljava/lang/ClassLoader;"
              "[Ldalvik/system/DexPathList$Element;"
              ")Ljava/lang/Object;"),
NATIVE_METHOD(DexFile, getClassLoaderContext,
              "(Ljava/lang/ClassLoader;"
              "[Ldalvik/system/DexPathList$Element;"
              ")Ljava/lang/String;"),
NATIVE_METHOD(DexFile, verifyInBackgroundNative,
              "(Ljava/lang/Object;"
              "Ljava/lang/ClassLoader;"
              "Ljava/lang/String;"
              ")V"),
NATIVE_METHOD(DexFile, isValidCompilerFilter, "(Ljava/lang/String;)Z"),
NATIVE_METHOD(DexFile, isProfileGuidedCompilerFilter,
              "(Ljava/lang/String;)Z"),
NATIVE_METHOD(DexFile, getNonProfileGuidedCompilerFilter,
              "(Ljava/lang/String;)Ljava/lang/String;"),
NATIVE_METHOD(DexFile, getSafeModeCompilerFilter,
              "(Ljava/lang/String;)Ljava/lang/String;"),
NATIVE_METHOD(DexFile, isBackedByOatFile, "(Ljava/lang/Object;)Z"),
NATIVE_METHOD(DexFile, getDexFileStatus,
    "(Ljava/lang/String;Ljava/lang/String;)Ljava/lang/String;"),
NATIVE_METHOD(DexFile, getDexFileOutputPaths,
    "(Ljava/lang/String;Ljava/lang/String;)[Ljava/lang/String;"),
NATIVE_METHOD(DexFile, getStaticSizeOfDexFile,
              "(Ljava/lang/Object;)J"),
NATIVE_METHOD(DexFile, getDexFileOptimizationStatus,
```

```
        "(Ljava/lang/String;Ljava/lang/String;)[Ljava/lang/String;"),
    NATIVE_METHOD(DexFile, setTrusted, "(Ljava/lang/Object;)V")
};
```

gMethods 是一个以 JNINativeMethod 为元素的数组，数组里面的具体内容则是 NATIVE_METHOD 宏及其所包含的参数。NATIVE_METHOD 宏的定义位于 libnativehelper/platform_include/nativehelper/jni_macros.h 中，代码如下：

```
//第 5 章/jni_macros.h
#define NATIVE_METHOD(className, functionName, signature)        \
    MAKE_JNI_NATIVE_METHOD(#functionName, signature,             \
                           className##_##functionName)
```

NATIVE_METHOD 宏继续又向下调用了几层宏的相关内容，代码如下：

```
//第 5 章/jni_macros.h
#define MAKE_JNI_NATIVE_METHOD(name, signature, function)                      \
    _NATIVEHELPER_JNI_MAKE_METHOD(kNormalNative, name, signature, function)

#define _NATIVEHELPER_JNI_MAKE_METHOD(kind, name, sig, fn)                     \
    MAKE_CHECKED_JNI_NATIVE_METHOD(kind, name, sig, fn)
```

最终，调用了 MAKE_CHECKED_JNI_NATIVE_METHOD 宏，它的实现位于 libnativehelper/platform_include/nativehelper/details/signature_checker.h 中，代码如下：

```
//第 5 章/signature_checker.h
//Expression to return JNINativeMethod, performs checking on
//signature + fn
#define MAKE_CHECKED_JNI_NATIVE_METHOD(native_kind, name_, signature_, fn)   \
    ([]() {                                                                   \
        using namespace nativehelper::detail;                                 \
        static_assert(                                                        \
            MatchJniDescriptorWithFunctionType<native_kind,                  \
                decltype(fn),                                                 \
                fn,                                                           \
                sizeof(signature_)>(signature_),                              \
            "JNI signature doesn't match C++ function type."); \
        /* Suppress implicit cast warnings by explicitly casting. */          \
        return JNINativeMethod {                                              \
            const_cast<decltype(JNINativeMethod::name)>(name_),              \
            const_cast<decltype(JNINativeMethod::signature)>(signature_),    \
            reinterpret_cast<void*>(&(fn))}; \
    })()
```

MAKE_CHECKED_JNI_NATIVE_METHOD 宏最终构建了一个完整的 JNINativeMethod。以 NATIVE_METHOD(DexFile, closeDexFile, "(Ljava/lang/Object;)Z")为例，它所构建出的 JNINativeMethod 中的 name 是 closeDexFile，signature 是"(Ljava/lang/Object;)Z"，fnPtr 指向的是 DexFile_closeDexFile()函数，该函数的实现也位于前文提到的 art/Runtime/native/dalvik_system_DexFile.cc 中。

此时，可以返回 JNI 的 RegisterNatives()函数中，构建 ArtMethod 并进行相关的验证，最终通过 m−>RegisterNative(fnPtr)语句调用了 ArtMethod 的 RegisterNative()函数。此处的 ArtMethod 变量 m 已经具有了 JNINativeMethod 中的 name 和 signature，并且给 ArtMethod 的 RegisterNative()函数传递 JNINativeMethod 中的 fnPtr。ArtMethod 的 RegisterNative()函数实现位于 art/Runtime/art_method.cc 中，代码如下：

```
//第5章/art_method.cc
const void* ArtMethod::RegisterNative(const void* native_method) {
  CHECK(IsNative()) << PrettyMethod();
  CHECK(native_method != nullptr) << PrettyMethod();
  void* new_native_method = nullptr;
  Runtime::Current()->GetRuntimeCallbacks()->RegisterNativeMethod(this,
      native_method, /* out */&new_native_method);
  SetEntryPointFromJni(new_native_method);
  return new_native_method;
}
```

这里通过调用 Runtime::Current()−>GetRuntimeCallbacks()函数先获取 RuntimeCallbacks，然后通过 RuntimeCallbacks::RegisterNativeMethod()函数来注册并获取新的本地方法 new_native_method，最后通过 SetEntryPointFromJni()函数设置函数入口。RuntimeCallbacks::RegisterNativeMethod()函数的实现位于 art/Runtime/Runtime_callbacks.cc 中，代码如下：

```
//第5章/Runtime_callbacks.cc
void RuntimeCallbacks::RegisterNativeMethod(ArtMethod* method,
    const void* in_cur_method, /* out */void** new_method) {
  void* cur_method = const_cast<void*>(in_cur_method);
  *new_method = cur_method;
  for (MethodCallback* cb : COPY(method_callbacks_)) {
    cb->RegisterNativeMethod(method, cur_method, new_method);
    if (*new_method != nullptr) {
      cur_method = *new_method;
    }
  }
}
```

这里通过遍历 RuntimeCallbacks 的 method_callbacks_ 中的元素，进而调用每个

MethodCallback 的 RegisterNativeMethod()函数。method_callbacks_ 是 std::vector<MethodCallback*>类型，所以其内部元素是 MethodCallback。MethodCallback 的 RegisterNativeMethod()函数在 art/Runtime/art_method.h 中被声明，代码如下：

```
//第5章/art_method.h
class MethodCallback {
public:
  virtual ~MethodCallback() {}

  virtual void RegisterNativeMethod(ArtMethod* method,
                                    const void* original_implementation,
                                    /*out*/void** new_implementation)
      REQUIRES_SHARED(Locks::mutator_lock_) = 0;
};
```

这里的 MethodCallback::RegisterNativeMethod()函数并没有实现。位于 art/openJDKJVMti/ti_method.cc 中的结构体 TiMethodCallback 继承了 MethodCallback，并实现了 RegisterNativeMethod()函数，代码如下：

```
//第5章/ti_method.cc
struct TiMethodCallback : public art::MethodCallback {
  void RegisterNativeMethod(art::ArtMethod* method,
                            const void* cur_method,
                            /*out*/void** new_method)
      override REQUIRES_SHARED(art::Locks::mutator_lock_) {
    if (event_handler->IsEventEnabledAnywhere(
        ArtJVMtiEvent::kNativeMethodBind)) {
      art::Thread* thread = art::Thread::Current();
      art::JNIEnvExt* jnienv = thread->GetJniEnv();
      ScopedLocalRef<jthread> thread_jni(
          jnienv, PhaseUtil::IsLivePhase() ?
          jnienv->AddLocalReference<jthread>(thread->GetPeer())
          : nullptr);
      art::ScopedThreadSuspension sts(thread, art::ThreadState::kNative);
      event_handler->DispatchEvent<ArtJVMtiEvent::kNativeMethodBind>(
          thread,
          static_cast<JNIEnv*>(jnienv),
          thread_jni.get(),
          art::jni::EncodeArtMethod(method),
          const_cast<void*>(cur_method),
          new_method);
    }
  }

  EventHandler* event_handler = nullptr;
};
```

这里通过 EventHandler 的 DispatchEvent() 函数，分发 ArtJVMtiEvent::kNativeMethodBind 本地方法绑定事件。EventHandler 的 DispatchEvent() 函数分发 ArtJVMtiEvent::kNativeMethodBind 的实现位于 art/openJDKJVMti/events-inl.h 文件中，代码如下：

```
//第5章/events-inl.h
//Need to give a custom specialization for NativeMethodBind since it has
//to deal with an out variable
template <>
inline void EventHandler::DispatchEvent < ArtJVMtiEvent::kNativeMethodBind >(
    art::Thread* thread, JNIEnv* jnienv, jthread jni_thread,
    jmethodID method, void* cur_method, void** new_method) const {
  art::ScopedThreadStateChange stsc(thread, art::ThreadState::kNative);
  std::vector < impl::EventHandlerFunc < ArtJVMtiEvent::kNativeMethodBind >> events =
      CollectEvents < ArtJVMtiEvent::kNativeMethodBind >(
          thread, jnienv, jni_thread, method, cur_method, new_method);
  *new_method = cur_method;
  for (auto event : events) {
    *new_method = cur_method;
    ExecuteCallback < ArtJVMtiEvent::kNativeMethodBind >(event,
                                                        jnienv,
                                                        jni_thread,
                                                        method,
                                                        cur_method,
                                                        new_method);
    if (*new_method != nullptr) {
      cur_method = *new_method;
    }
  }
  *new_method = cur_method;
}
```

这里的 ExecuteCallback 最终通过宏的展开执行了 NativeMethodBind，而 NativeMethodBind 这个事件是被 JVMti-agents 监听的，所以在 art/tools/JVMti-agents/tifast/tifast.cc 中，NativeMethodBind 被对应到 NATIVE_METHOD_BIND，然后又被对应到 can_generate_native_method_bind_events，代码如下：

```
//第5章/tifast.cc
    fun(NativeMethodBind, EVENT(NATIVE_METHOD_BIND), (JVMtiEnv* JVMti,
    JNIEnv* jni, jthread thread, jmethodID meth, void* v1, void** v2),
    (JVMti, jni, jthreadContainer{.thread = thread}, meth, v1, v2))
    ...
    DO_CASE(NATIVE_METHOD_BIND, can_generate_native_method_bind_events);
```

can_generate_native_method_bind_events 最后的处理位于 art/openJDKJVMti/OpenJDKJVMTi.cc 中。

至此，基本完成了对于运行时本地方法初始化的一个初步介绍，对于后续内容可以根据需求进一步深入分析代码。

5.6 ART 启动中的其他本地方法的注册

ART 启动过程中所涉及的本地方法的注册，除了上文提到的运行时本地方法的注册，还有 Android 本地方法的注册。本部分内容将对 Android 本地方法的注册进行分析。

虚拟机启动之后，沿着代码的调用流程，可以向上追溯到 frameworks/base/core/jni/AndroidRuntime.cpp 文件中的 AndroidRuntime::start()函数中，在该函数中可以看到，启动了虚拟机之后，会对本地方法进行注册，它也是 ART 启动的一部分。

这里需要说明的是，前文所介绍的 Runtime::Start()函数(art/Runtime/Runtime.cc)中的本地方法的注册，主要是注册运行时本地方法，而本部分注册的本地方法则是 Android 整个框架的本地方法，这个部分的注册发生在虚拟机启动之后，而 Runtime::Start()函数(art/Runtime/Runtime.cc)中的本地方法的注册发生在虚拟机启动阶段，二者有一个先后顺序。在后续的代码分析中也可以根据注册方法的名字很显然地做出区分。

上面已经追踪到了 AndroidRuntime::start()函数，代码如下：

```
//第 5 章/ Runtime.cc
/*
 * Start the Android Runtime. This involves starting the virtual machine
 * and calling the "static void main(String[] args)" method in the class
 * named by "className".
 *
 * Passes the main function two arguments, the class name and the
 * specified options string.
 */
void AndroidRuntime::start(
    const char* className, const Vector<String8>& options,
    bool zygote) {
    …
    /* start the virtual machine */
    JniInvocation jni_invocation;
    jni_invocation.Init(NULL);
    JNIEnv* env;
    if (startVm(&mJavaVM, &env, zygote) != 0) {
        return;
    }
    onVmCreated(env);
```

```
    /*
     * Register android functions.
     */
    if (startReg(env) < 0) {
        ALOGE("Unable to register all android natives\n");
        return;
    }
    ...
}
```

其中，startReg()函数所进行的是注册本地方法的工作，其注释描述的是"Register android functions."，可以理解为注册 Android 函数，其失败的 log 信息 Unable to register all android natives 表示不能注册所有的 Android 本地方法，所以可以判断其作用是注册 Android 体系的本地方法。startReg()函数的实现位于 frameworks/base/core/jni/AndroidRuntime.cpp 文件中，代码如下：

```
//第5章/AndroidRuntime.cpp
/*
 * Register android native functions with the VM.
 */
/*static*/ int AndroidRuntime::startReg(JNIEnv* env)
{
    ATRACE_NAME("RegisterAndroidNatives");
    /*
     * This hook causes all future threads created in this process to be
     * attached to the JavaVM.  (This needs to go away in favor of JNI
     * Attach calls.)
     */
    androidSetCreateThreadFunc((android_create_thread_fn) javaCreateThreadEtc);

    ALOGV("--- registering native functions ---\n");

    /*
     * Every "register" function calls one or more things that return
     * a local reference (e.g. FindClass).  Because we haven't really
     * started the VM yet, they're all getting stored in the base frame
     * and never released. Use Push/Pop to manage the storage.
     */
    env->PushLocalFrame(200);

    if (register_jni_procs(gRegJNI, NELEM(gRegJNI), env) < 0) {
        env->PopLocalFrame(NULL);
        return -1;
    }
```

```
        env->PopLocalFrame(NULL);

        //createJavaThread("fubar", quickTest, (void*) "hello");

        return 0;
}
```

startReg()函数的注释说得很清楚,它的作用就是要"Register android native functions with the VM.",即向 VM 注册 Android 的本地方法,并且,"Because we haven't really started the VM yet, they're all getting stored in the base frame and never released. Use Push/Pop to manage the storage."这里说的虚拟机并没有真正地被启动,指的是 Zygote 进程的初始化还没完成,等 Zygote 进程的初始化完成之后,才认为真正地完成了 VM 的启动,因为 Zygote 进程的初始化完成之后,才能开始接受 App 的请求,为其分配新的进程。

在 startReg()函数中调用了一个重要的函数 register_jni_procs()。register_jni_procs()函数的实现同样位于 frameworks/base/core/jni/AndroidRuntime.cpp 中,代码如下:

```
//第 5 章/AndroidRuntime.cpp
static int register_jni_procs(
    const RegJNIRec array[], size_t count, JNIEnv* env) {
    for (size_t i = 0; i < count; i++) {
        if (array[i].mProc(env) < 0) {
#ifndef NDEBug
            ALOGD("----------!!! %s failed to load\n", array[i].mName);
#endif
            return -1;
        }
    }
    return 0;
}
```

其中,register_jni_procs()函数接收的第 1 个参数是 gRegJNI,它是一个数组,register_jni_procs()函数逐个遍历这个数组的每个元素,然后执行这个元素的 mProc()函数。gRegJNI 作为一个数组,它的代码如下:

```
//第 5 章/AndroidRuntime.cpp
static const RegJNIRec gRegJNI[] = {
    REG_JNI(register_com_android_internal_os_RuntimeInit),
    REG_JNI(register_com_android_internal_os_ZygoteInit_nativeZygoteInit),
    REG_JNI(register_android_os_SystemClock),
    REG_JNI(register_android_util_EventLog),
    REG_JNI(register_android_util_Log),
    REG_JNI(register_android_util_MemoryIntArray),
    REG_JNI(register_android_util_PathParser),
```

```cpp
    REG_JNI(register_android_util_StatsLog),
    REG_JNI(register_android_util_StatsLogInternal),
    REG_JNI(register_android_app_admin_SecurityLog),
    REG_JNI(register_android_content_AssetManager),
    REG_JNI(register_android_content_StringBlock),
    REG_JNI(register_android_content_XmlBlock),
    REG_JNI(register_android_content_res_ApkAssets),
    REG_JNI(register_android_text_AndroidCharacter),
    REG_JNI(register_android_text_Hyphenator),
    REG_JNI(register_android_view_InputDevice),
    REG_JNI(register_android_view_KeyCharacterMap),
    REG_JNI(register_android_os_Process),
    REG_JNI(register_android_os_SystemProperties),
    REG_JNI(register_android_os_Binder),
    REG_JNI(register_android_os_Parcel),
    REG_JNI(register_android_os_HidlSupport),
    REG_JNI(register_android_os_HwBinder),
    REG_JNI(register_android_os_HwBlob),
    REG_JNI(register_android_os_HwParcel),
    REG_JNI(register_android_os_HwRemoteBinder),
    REG_JNI(register_android_os_NativeHandle),
    REG_JNI(register_android_os_VintfObject),
    REG_JNI(register_android_os_VintfRuntimeInfo),
    REG_JNI(register_android_graphics_Canvas),
    //This needs to be before register_android_graphics_Graphics, or the
    //latter will not be able to find the jmethodID for ColorSpace.get()
    REG_JNI(register_android_graphics_ColorSpace),
    REG_JNI(register_android_graphics_Graphics),
    REG_JNI(register_android_view_DisplayEventReceiver),
    REG_JNI(register_android_view_RenderNode),
    REG_JNI(register_android_view_RenderNodeAnimator),
    REG_JNI(register_android_view_DisplayListCanvas),
    REG_JNI(register_android_view_InputApplicationHandle),
    REG_JNI(register_android_view_InputWindowHandle),
    REG_JNI(register_android_view_TextureLayer),
    REG_JNI(register_android_view_ThreadedRenderer),
    REG_JNI(register_android_view_Surface),
    REG_JNI(register_android_view_SurfaceControl),
    REG_JNI(register_android_view_SurfaceSession),
    REG_JNI(register_android_view_CompositionSamplingListener),
    REG_JNI(register_android_view_TextureView),

    REG_JNI(register_com_android_internal_view_animation_NativeInterpolatorFactoryHelper),
    REG_JNI(register_com_google_android_gles_jni_EGLImpl),
    REG_JNI(register_com_google_android_gles_jni_GLImpl),
    REG_JNI(register_android_opengl_jni_EGL14),
    REG_JNI(register_android_opengl_jni_EGL15),
```

```cpp
REG_JNI(register_android_opengl_jni_EGLExt),
REG_JNI(register_android_opengl_jni_GLES10),
REG_JNI(register_android_opengl_jni_GLES10Ext),
REG_JNI(register_android_opengl_jni_GLES11),
REG_JNI(register_android_opengl_jni_GLES11Ext),
REG_JNI(register_android_opengl_jni_GLES20),
REG_JNI(register_android_opengl_jni_GLES30),
REG_JNI(register_android_opengl_jni_GLES31),
REG_JNI(register_android_opengl_jni_GLES31Ext),
REG_JNI(register_android_opengl_jni_GLES32),

REG_JNI(register_android_graphics_Bitmap),
REG_JNI(register_android_graphics_BitmapFactory),
REG_JNI(register_android_graphics_BitmapRegionDecoder),
REG_JNI(register_android_graphics_ByteBufferStreamAdaptor),
REG_JNI(register_android_graphics_Camera),
REG_JNI(register_android_graphics_CreateJavaOutputStreamAdaptor),
REG_JNI(register_android_graphics_CanvasProperty),
REG_JNI(register_android_graphics_ColorFilter),
REG_JNI(register_android_graphics_DrawFilter),
REG_JNI(register_android_graphics_FontFamily),
REG_JNI(register_android_graphics_GraphicBuffer),
REG_JNI(register_android_graphics_ImageDecoder),
REG_JNI(register_android_graphics_drawable_AnimatedImageDrawable),
REG_JNI(register_android_graphics_Interpolator),
REG_JNI(register_android_graphics_MaskFilter),
REG_JNI(register_android_graphics_Matrix),
REG_JNI(register_android_graphics_Movie),
REG_JNI(register_android_graphics_NinePatch),
REG_JNI(register_android_graphics_Paint),
REG_JNI(register_android_graphics_Path),
REG_JNI(register_android_graphics_PathMeasure),
REG_JNI(register_android_graphics_PathEffect),
REG_JNI(register_android_graphics_Picture),
REG_JNI(register_android_graphics_Region),
REG_JNI(register_android_graphics_Shader),
REG_JNI(register_android_graphics_SurfaceTexture),
REG_JNI(register_android_graphics_Typeface),
REG_JNI(register_android_graphics_YuvImage),
REG_JNI(register_android_graphics_drawable_AnimatedVectorDrawable),
REG_JNI(register_android_graphics_drawable_VectorDrawable),
REG_JNI(register_android_graphics_fonts_Font),
REG_JNI(register_android_graphics_fonts_FontFamily),
REG_JNI(register_android_graphics_pdf_PdfDocument),
REG_JNI(register_android_graphics_pdf_PdfEditor),
REG_JNI(register_android_graphics_pdf_PdfRenderer),
```

```
REG_JNI(register_android_graphics_text_MeasuredText),
REG_JNI(register_android_graphics_text_LineBreaker),

REG_JNI(register_android_database_CursorWindow),
REG_JNI(register_android_database_SQLiteConnection),
REG_JNI(register_android_database_SQLiteGlobal),
REG_JNI(register_android_database_SQLiteDebug),
REG_JNI(register_android_os_Debug),
REG_JNI(register_android_os_FileObserver),
REG_JNI(register_android_os_GraphicsEnvironment),
REG_JNI(register_android_os_MessageQueue),
REG_JNI(register_android_os_SELinux),
REG_JNI(register_android_os_Trace),
REG_JNI(register_android_os_UEventObserver),
REG_JNI(register_android_net_LocalSocketImpl),
REG_JNI(register_android_net_NetworkUtils),
REG_JNI(register_android_os_MemoryFile),
REG_JNI(register_android_os_SharedMemory),
REG_JNI(register_com_android_internal_os_ClassLoaderFactory),
REG_JNI(register_com_android_internal_os_Zygote),
REG_JNI(register_com_android_internal_os_ZygoteInit),
REG_JNI(register_com_android_internal_util_VirtualRefBasePtr),
REG_JNI(register_android_hardware_Camera),
REG_JNI(register_android_hardware_camera2_CameraMetadata),
REG_JNI(register_android_hardware_camera2_legacy_LegacyCameraDevice),
REG_JNI(register_android_hardware_camera2_legacy_PerfMeasurement),
REG_JNI(register_android_hardware_camera2_DngCreator),
REG_JNI(register_android_hardware_HardwareBuffer),
REG_JNI(register_android_hardware_SensorManager),
REG_JNI(register_android_hardware_SerialPort),
REG_JNI(register_android_hardware_SoundTrigger),
REG_JNI(register_android_hardware_UsbDevice),
REG_JNI(register_android_hardware_UsbDeviceConnection),
REG_JNI(register_android_hardware_UsbRequest),

REG_JNI(register_android_hardware_location_ActivityRecognitionHardware),
REG_JNI(register_android_media_AudioEffectDescriptor),
REG_JNI(register_android_media_AudioSystem),
REG_JNI(register_android_media_AudioRecord),
REG_JNI(register_android_media_AudioTrack),
REG_JNI(register_android_media_AudioAttributes),
REG_JNI(register_android_media_AudioProductStrategies),
REG_JNI(register_android_media_AudioVolumeGroups),
REG_JNI(register_android_media_AudioVolumeGroupChangeHandler),
REG_JNI(register_android_media_JetPlayer),
REG_JNI(register_android_media_MicrophoneInfo),
REG_JNI(register_android_media_RemoteDisplay),
```

```
        REG_JNI(register_android_media_ToneGenerator),
        REG_JNI(register_android_media_midi),

        REG_JNI(register_android_opengl_classes),
        REG_JNI(register_android_server_NetworkManagementSocketTagger),
        REG_JNI(register_android_ddm_DdmHandleNativeHeap),
        REG_JNI(register_android_backup_BackupDataInput),
        REG_JNI(register_android_backup_BackupDataOutput),
        REG_JNI(register_android_backup_FileBackupHelperBase),
        REG_JNI(register_android_backup_BackupHelperDispatcher),
        REG_JNI(register_android_app_backup_FullBackup),
        REG_JNI(register_android_app_Activity),
        REG_JNI(register_android_app_ActivityThread),
        REG_JNI(register_android_app_NativeActivity),
        REG_JNI(register_android_util_jar_StrictJarFile),
        REG_JNI(register_android_view_InputChannel),
        REG_JNI(register_android_view_InputEventReceiver),
        REG_JNI(register_android_view_InputEventSender),
        REG_JNI(register_android_view_InputQueue),
        REG_JNI(register_android_view_KeyEvent),
        REG_JNI(register_android_view_MotionEvent),
        REG_JNI(register_android_view_PointerIcon),
        REG_JNI(register_android_view_VelocityTracker),

        REG_JNI(register_android_content_res_ObbScanner),
        REG_JNI(register_android_content_res_Configuration),

        REG_JNI(register_android_animation_PropertyValuesHolder),
        REG_JNI(register_android_security_Scrypt),
        REG_JNI(register_com_android_internal_content_NativeLibraryHelper),
        REG_JNI(register_com_android_internal_os_AtomicDirectory),
        REG_JNI(register_com_android_internal_os_FuseAppLoop),
};
```

从这里要注册的本地方法的名称可以看到其与运行时的本地方法的差别。其中，RegJNIRec 及 REG_JNI 的定义如下：

```
//第 5 章/AndroidRuntime.cpp
#ifdef NDEBug
    #define REG_JNI(name)         { name }
    struct RegJNIRec {
        int ( *mProc)(JNIEnv * );
    };
#else
    #define REG_JNI(name)         { name, #name }
```

```
    struct RegJNIRec {
        int ( * mProc)(JNIEnv * );
        const char * mName;
    };
#endif
```

从上述代码可以知道，在非调试情况下，RegJNIRec 结构体中只有一个函数指针，这个函数指针指向了 REG_JNI(name) 中与 name 同名的函数，所以我们可以在 frameworks/base/core/jni/AndroidRuntime.cpp 中找到所有 REG_JNI(name) 中 name 所对应的 register_xxxxxx 开头的函数，所以前面提到的 register_jni_procs() 函数等于将 gRegJNI 数组中所有 REG_JNI(name) 中与 name 同名的函数全部执行了一遍。

这些函数有的是直接在当前文件中实现的，例如 register_com_android_internal_os_ZygoteInit_nativeZygoteInit 函数，它是 gRegJNI 数组的第二项，代码如下：

```
//第 5 章/AndroidRuntime.cpp
int register_com_android_internal_os_ZygoteInit_nativeZygoteInit(
    JNIEnv * env) {
    const JNINativeMethod methods[] = {
        { "nativeZygoteInit", "()V",
        (void * ) com_android_internal_os_ZygoteInit_nativeZygoteInit },
    };
    return jniRegisterNativeMethods(env, "com/android/internal/os/ZygoteInit",
                                    methods, NELEM(methods));
}
```

register_com_android_internal_os_ZygoteInit_nativeZygoteInit() 函数最终被转化为对 jniRegisterNativeMethods() 函数的调用，并且传递的参数也是 env、类名、methods 列表和 methods 中的 method 数量，这几个参数在运行时本地方法的注册中已经介绍过，这里也是类似的情况。此外，还有一批函数的实现不在当前文件之下，所以进行了外部声明，例如 register_android_os_SystemClock() 函数，它是 gRegJNI 数组的第三项，所以它被声明为外部实现，代码如下：

```
extern int register_android_os_SystemClock(JNIEnv * env);
```

register_android_os_SystemClock() 函数的实现实际上位于 frameworks/base/core/jni/android_os_SystemClock.cpp 中，代码如下：

```
//第 5 章/android_os_SystemClock.cpp
int register_android_os_SystemClock(JNIEnv * env) {
    return RegisterMethodsOrDie(env, "android/os/SystemClock", gMethods, NELEM(gMethods));
}
```

这类声明为外部实现的函数都能找到一个与之对应的除去前缀 register_ 的同名文件，在其中可以找到具体的函数实现。register_android_os_SystemClock() 函数最终通过了 frameworks/base/core/jni/core_jni_helpers.h 中的 RegisterMethodsOrDie() 函数调用了 frameworks/base/core/jni/AndroidRuntime.cpp 中的 AndroidRuntime::registerNativeMethods() 函数。RegisterMethodsOrDie() 函数的代码如下：

```
//第 5 章/core_jni_helpers.h
static inline int RegisterMethodsOrDie(
        JNIEnv* env, const char* className,
        const JNINativeMethod* gMethods, int numMethods)
{
    int res = AndroidRuntime::registerNativeMethods(env, className,
            gMethods, numMethods);
    LOG_ALWAYS_FATAL_IF(res < 0, "Unable to register native methods.");
    return res;
}
```

而 AndroidRuntime::registerNativeMethods() 函数最终也是调用了 jniRegisterNativeMethods() 函数，代码如下：

```
//第 5 章/AndroidRuntime.cpp
/*
 * Register native methods using JNI.
 */
/*static*/ int AndroidRuntime::registerNativeMethods(JNIEnv* env,
    const char* className, const JNINativeMethod* gMethods,
    int numMethods) {
    return jniRegisterNativeMethods(env, className, gMethods,
                                    numMethods);
}
```

这类与 register_android_os_SystemClock() 函数一样的外部声明函数，都采用了类似的处理流程，所以所有的本地方法注册最终都会调用 jniRegisterNativeMethods() 函数。jniRegisterNativeMethods() 函数的实现位于 libnativehelper/JNIHelp.cpp 中，代码如下：

```
//第 5 章/JNIHelp.cpp
MODULE_API int jniRegisterNativeMethods(C_JNIEnv* env,
    const char* className, const JNINativeMethod* gMethods,
    int numMethods)
{
    JNIEnv* e = reinterpret_cast<JNIEnv*>(env);

    ALOGV("Registering %s's %d native methods...", className, numMethods);
```

```
    scoped_local_ref<jclass> c(env, findClass(env, className));
    ALOG_ALWAYS_FATAL_IF(c.get() == NULL,
          "Native registration unable to find class '%s'; aborting...",
          className);

    int result = e->RegisterNatives(c.get(), gMethods, numMethods);
    ALOG_ALWAYS_FATAL_IF(result < 0, "RegisterNatives failed for '%s'; aborting...",
className);

    return 0;
}
```

这里最终通过调用 JNIEnv 的 RegisterNatives() 函数完成了本地方法的注册,而 JNIEnv 的 RegisterNatives() 函数经过转换,最终调用的是 JNI::RegisterNatives() 函数,这个转换过程在 ART 启动中的守护线程部分中有过介绍,此处不再重复介绍。JNI::RegisterNatives() 函数的实现位于 art/Runtime/jni/jni_internal.cc 中,往下的内容就和注册运行时的本地方法一样了。

对比观察注册运行时本地方法和本部分的本地方法注册,其核心内容和思想都是一致的,所不同的是在前期各个模块怎么去处理相关内容及准备 JNINativeMethod * gMethods 中所需要的 name、signatureh 的内容。将这二者对比也会发现,整体的本地方法的注册采用的还是同样的模式。

5.7 Zygote 进程

ART 的启动,开始于 Zygote 进程的启动函数中,ART 可以看作 Zygote 进程中所做的一部分工作。本章前面的内容介绍了与 ART 启动相关的内容,都属于 Zygote 进程。此外,还有 Zygote 进程自身及与进程启动相关的代码并未进行介绍,本部分内容将对此进行介绍。

Zygote 进程作为 Android 的一个非常重要的进程,它是通过 Android 系统的 init 进程创建的,init 进程是 Android 系统的第 1 个进程,系统内的其他进程都可以看作它的子孙进程。这是因为 Android 系统采用的是 Linux 内核,有关 init 进程及其创建子孙进程所采用的 fork 机制等都和 Linux 系统类似。Zygote 进程负责 ART 中各项要素的启动,并且负责对 System Server 进程及应用进程的创建。理清楚 Zygote 进程启动的相关内容,有利于理解 ART 启动的位置,同时也能搞清楚 App 在 ART 上持续不断地运行的一些相关内容。

Zygote 进程启动的 main() 函数为 frameworks/base/cmds/app_process/app_main.cpp 中的 main() 函数,从 main() 函数往下跟踪,可以跟踪到 AndroidRuntime 的 start() 函数,它的实现位于 rameworks/base/core/jni/AndroidRuntime.cpp 中。这个流程在 ART 启动中进行过介绍,在 AndroidRuntime 的 start() 函数中,除了启动了虚拟机,还剩余一部

分代码没进行分析,它是 Zygote 进程启动本身的代码,也是 App 源源不断地被送到 ART 去执行的核心内容,这部分的代码如下:

```cpp
//第 5 章/AndroidRuntime.cpp
/*
 * Start the Android Runtime. This involves starting the virtual machine
 * and calling the "static void main(String[] args)" method in the class
 * named by "className".
 *
 * Passes the main function two arguments, the class name and the
 * specified * options string.
 */
void AndroidRuntime::start(
    const char * className, const Vector<String8> & options,
    bool zygote) {
…
    char * slashClassName = toSlashClassName(className != NULL ?
                                                className : "");
    jclass startClass = env->FindClass(slashClassName);
    if (startClass == NULL) {
        ALOGE("JavaVM unable to locate class '%s'\n", slashClassName);
        /* keep going */
    } else {
        jmethodID startMeth = env->GetStaticMethodID(startClass, "main",
            "([Ljava/lang/String;)V");
        if (startMeth == NULL) {
            ALOGE("JavaVM unable to find main() in '%s'\n", className);
            /* keep going */
        } else {
            env->CallStaticVoidMethod(startClass, startMeth, strArray);

#if 0
            if (env->ExceptionCheck())
                threadExitUncaughtException(env);
#endif
        }
    }
    free(slashClassName);

    ALOGD("Shutting down VM\n");
    if (mJavaVM->DetachCurrentThread() != JNI_OK)
        ALOGW("Warning: unable to detach main thread\n");
    if (mJavaVM->DestroyJavaVM() != 0)
        ALOGW("Warning: VM did not shut down cleanly\n");
}
```

其中,核心的代码是 CallStaticVoidMethod() 函数,它会通过 JNI 接口调用 com. android.internal.os.ZygoteInit 类的 main() 函数,进行 Zygote 初始化的所有工作。com. android.internal.os.ZygoteInit 类的实现位于 frameworks/base/core/java/com/android/ internal/os/ZygoteInit.java 中,main() 函数的代码如下:

```java
//第5章/ZygoteInit.java
public static void main(String argv[]) {
    ZygoteServer zygoteServer = null;
    
    //Mark zygote start. This ensures that thread creation will throw
    //an error.
    ZygoteHooks.startZygoteNoThreadCreation();
    
    //Zygote goes into its own process group.
    try {
        Os.setpgid(0, 0);
    } catch (ErrnoException ex) {
        throw new RuntimeException("Failed to setpgid(0,0)", ex);
    }
    
    Runnable caller;
    try {
        //Report Zygote start time to tron unless it is a Runtime
        //restart
        if (!"1".equals(SystemProperties.get("sys.boot_completed"))) {
            MetricsLogger.histogram(null, "boot_zygote_init",
                              (int) SystemClock.elapsedRealtime());
        }
        
        String bootTimeTag = Process.is64Bit() ? "Zygote64Timing" :
                                                 "Zygote32Timing";
        TimingsTraceLog bootTimingsTraceLog = new
                TimingsTraceLog(bootTimeTag, Trace.TRACE_TAG_DALVIK);
        bootTimingsTraceLog.traceBegin("ZygoteInit");
        RuntimeInit.enableDdms();
        
        boolean startSystemServer = false;
        String zygoteSocketName = "zygote";
        String abiList = null;
        boolean enableLazyPreload = false;
        for (int i = 1; i < argv.length; i++) {
            if ("start-system-server".equals(argv[i])) {
                startSystemServer = true;
            } else if ("--enable-lazy-preload".equals(argv[i])) {
                enableLazyPreload = true;
```

```java
            } else if (argv[i].startsWith(ABI_LIST_ARG)) {
                abiList = argv[i].substring(ABI_LIST_ARG.length());
            } else if (argv[i].startsWith(SOCKET_NAME_ARG)) {
                zygoteSocketName = argv[i].substring(
                        SOCKET_NAME_ARG.length());
            } else {
                throw new RuntimeException(
                        "Unknown command line argument: " + argv[i]);
            }
        }

        final boolean isPrimaryZygote = zygoteSocketName.equals(
Zygote.PRIMARY_SOCKET_NAME);

        if (abiList == null) {
            throw new RuntimeException("No ABI list supplied.");
        }

        //In some configurations, we avoid preloading resources and
        //classes eagerly.
        //In such cases, we will preload things prior to our first fork
        if (!enableLazyPreload) {
            bootTimingsTraceLog.traceBegin("ZygotePreload");
            EventLog.writeEvent(LOG_BOOT_PROGRESS_PRELOAD_START,
                           SystemClock.uptimeMillis());
            preload(bootTimingsTraceLog);
            EventLog.writeEvent(LOG_BOOT_PROGRESS_PRELOAD_END,
                           SystemClock.uptimeMillis());
            bootTimingsTraceLog.traceEnd(); //ZygotePreload
        } else {
            Zygote.resetNicePriority();
        }

        //Do an initial gc to clean up after startup
        bootTimingsTraceLog.traceBegin("PostZygoteInitGC");
        gcAndFinalize();
        bootTimingsTraceLog.traceEnd(); //PostZygoteInitGC

        bootTimingsTraceLog.traceEnd(); //ZygoteInit
        //Disable tracing so that forked processes do not inherit stale
        //tracing tags from Zygote
        Trace.setTracingEnabled(false, 0);


        Zygote.initNativeState(isPrimaryZygote);
```

```java
        ZygoteHooks.stopZygoteNoThreadCreation();

        zygoteServer = new ZygoteServer(isPrimaryZygote);

        if (startSystemServer) {
            Runnable r = forkSystemServer(abiList, zygoteSocketName,
                                    zygoteServer);

            //{@code r == null} in the parent (zygote) process, and
            //{@code r != null} in the child (system_server) process
            if (r != null) {
                r.run();
                return;
            }
        }

        Log.i(TAG, "Accepting command socket connections");

        //The select loop returns early in the child process after a
        //fork and loops forever in the zygote
        caller = zygoteServer.runSelectLoop(abiList);
    } catch (Throwable ex) {
        Log.e(TAG, "System zygote died with exception", ex);
        throw ex;
    } finally {
        if (zygoteServer != null) {
            zygoteServer.closeServerSocket();
        }
    }

    //We're in the child process and have exited the select loop.
    //Proceed to execute the command
    if (caller != null) {
        caller.run();
    }
}
```

这里比较重要的函数是 ZygoteInit.forkSystemServer() 函数和 ZygoteServer.runSelectLoop() 函数,前者用来创建 SystemServer 进程,后者是不断的循环监听命令,为每个要运行的 App 创建新的进程,下面对这两个函数分别进行介绍。

5.7.1　System Server 进程

System Server 进程的创建位于 ZygoteInit 类的 forkSystemServer() 函数中,该函数用于准备参数并且 fork 了一个新进程,这个新进程就是 System Server,在运行的 Android 系统中,该进程通常叫作 system_server 进程。

fork 新的进程的操作并不在 forkSystemServer() 函数之内直接执行,而是通过调用 Zygote 类的 forkSystemServer() 函数。Zygote 类的 forkSystemServer() 函数会为新进程(子进程)返回值为 0 的 pid,然后 ZygoteInit 类的 forkSystemServer() 函数会为新进程执行 handleSystemServerProcess() 函数,并将其结果作为返回值,这时候这个返回值为非空,而 Zygote 类的 forkSystemServer() 函数会为原有进程(父进程)返回子进程的 pid,这时候 pid 不等于 0,所以会直接执行 ZygoteInit 的 forkSystemServer() 函数最后返回一个空值。ZygoteInit 的 forkSystemServer() 函数的代码如下:

```java
//第 5 章/ZygoteInit.java
/**
 * Prepare the arguments and forks for the system server process.
 *
 * @return A {@code Runnable} that provides an entrypoint into
 * system_server code in the child process; {@code null} in the parent.
 */
private static Runnable forkSystemServer(String abiList,
        String socketName, ZygoteServer zygoteServer) {
    long capabilities = posixCapabilitiesAsBits(
            OsConstants.CAP_IPC_LOCK,
            OsConstants.CAP_KILL,
            OsConstants.CAP_NET_ADMIN,
            OsConstants.CAP_NET_BIND_SERVICE,
            OsConstants.CAP_NET_BROADCAST,
            OsConstants.CAP_NET_RAW,
            OsConstants.CAP_SYS_MODULE,
            OsConstants.CAP_SYS_NICE,
            OsConstants.CAP_SYS_PTRACE,
            OsConstants.CAP_SYS_TIME,
            OsConstants.CAP_SYS_TTY_CONFIG,
            OsConstants.CAP_WAKE_ALARM,
            OsConstants.CAP_BLOCK_SUSPEND
    );
    /* Containers run without some capabilities, so drop any caps that

    are not available. */
    StructCapUserHeader header = new StructCapUserHeader(
            OsConstants._LINUX_CAPABILITY_VERSION_3, 0);
    StructCapUserData[] data;
    try {
        data = Os.capget(header);
    } catch (ErrnoException ex) {
        throw new RuntimeException("Failed to capget()", ex);
    }
```

```java
        capabilities &= ((long) data[0].effective) |
                (((long) data[1].effective) << 32);

        /* Hardcoded command line to start the system server */
        String args[] = {
                "--setuid=1000",
                "--setgid=1000",
                "--setgroups=1001,1002,1003,1004,1005,1006,1007,1008,1009,1010,1018,"
                + "1021,1023," + "1024,1032,1065,3001,3002,3003,3006,3007,3009,3010",
                "--capabilities=" + capabilities + "," + capabilities,
                "--nice-name=system_server",
                "--runtime-args",
                "--target-sdk-version=" +
                        VMRuntime.SDK_VERSION_CUR_DEVELOPMENT,
                "com.android.server.SystemServer",
        };
        ZygoteArguments parsedArgs = null;

        int pid;

        try {
            parsedArgs = new ZygoteArguments(args);
            Zygote.applyDebuggerSystemProperty(parsedArgs);
            Zygote.applyInvokeWithSystemProperty(parsedArgs);

            boolean profileSystemServer = SystemProperties.getBoolean(
                    "dalvik.vm.profilesystemserver", false);
            if (profileSystemServer) {
                parsedArgs.mRuntimeFlags |= Zygote.PROFILE_SYSTEM_SERVER;
            }

            /* Request to fork the system server process */
            pid = Zygote.forkSystemServer(
                    parsedArgs.mUid, parsedArgs.mGid,
                    parsedArgs.mGids,
                    parsedArgs.mRuntimeFlags,
                    null,
                    parsedArgs.mPermittedCapabilities,
                    parsedArgs.mEffectiveCapabilities);
        } catch (IllegalArgumentException ex) {
            throw new RuntimeException(ex);
        }

        /* For child process */
        if (pid == 0) {
            if (hasSecondZygote(abiList)) {
```

```
            waitForSecondaryZygote(socketName);
        }

        zygoteServer.closeServerSocket();
        return handleSystemServerProcess(parsedArgs);
    }

    return null;
}
```

ZygoteInit 的 forkSystemServer()函数调用了 Zygote 类的 forkSystemServer()函数。Zygote 类的 forkSystemServer() 函数位于 frameworks/base/core/java/com/android/internal/os/Zygote.java 中,代码如下:

```
//第5章/Zygote.java
/**
 * Special method to start the system server process. In addition to the
 * common actions performed in forkAndSpecialize, the pid of the child
 * process is recorded such that the death of the child process will
 * cause zygote to exit.
 *
 * @param uid the UNIX uid that the new process should setuid() to after
 * fork()ing and and before spawning any threads.
 * @param gid the UNIX gid that the new process should setgid() to after
 * fork()ing and and before spawning any threads.
 * @param gids null-ok; a list of UNIX gids that the new process should
 * setgroups() to after fork and before spawning any threads.
 * @param RuntimeFlags bit flags that enable ART features.
 * @param rlimits null-ok an array of rlimit tuples, with the second
 * dimension having a length of 3 and representing
 * (resource, rlim_cur, rlim_max). These are set via the posix
 * setrlimit(2) call.
 * @param permittedCapabilities argument for setcap()
 * @param effectiveCapabilities argument for setcap()
 *
 * @return 0 if this is the child, pid of the child
 * if this is the parent, or -1 on error.
 */
public static int forkSystemServer(int uid, int gid, int[] gids,
        int RuntimeFlags, int[][] rlimits, long permittedCapabilities,
        long effectiveCapabilities) {
    ZygoteHooks.preFork();
    //Resets nice priority for zygote process
    resetNicePriority();
    int pid = nativeForkSystemServer(
```

```
                uid, gid, gids, RuntimeFlags, rlimits,
                permittedCapabilities, effectiveCapabilities);
    //Enable tracing as soon as we enter the system_server
    if (pid == 0) {
        Trace.setTracingEnabled(true, RuntimeFlags);
    }
    ZygoteHooks.postForkCommon();
    return pid;
}
```

Zygote 类的 forkSystemServer() 函数通过 nativeForkSystemServer() 函数创建了一个新进程，nativeForkSystemServer() 函数并没有在 Java 类 Zygote 中实现，它是一个 JNI 的本地函数，最终实现是 frameworks/base/core/jni/com_android_internal_os_Zygote.cpp 中的 com_android_internal_os_Zygote_nativeForkSystemServer() 函数，而 com_android_internal_os_Zygote_nativeForkSystemServer() 函数通过调用同文件中的 ForkCommon() 函数实现了 fork 新进程的动作，ForkCommon() 函数最终调用了 fork() 函数实现了新进程的创建。fork() 函数在创建新进程的时候实现了一次调用两次返回，分别为父进程返回了子进程的进程标识符，并为子进程返回 0。在这种情况下，沿着调用关系回溯回去，Zygote 类的 forkSystemServer() 函数也有两次返回，分别为父进程返回子进程的进程符，并为子进程返回 0。

ZygoteInit 的 forkSystemServer() 函数获取的 Zygote 类的 forkSystemServer() 函数的返回值不为 0，说明目前进程是父进程，会直接为 ZygoteInit 的 forkSystemServer() 函数返回 null。否则，说明目前进程是新创建的子进程，会调用 ZygoteInit 类的 handleSystemServerProcess() 函数，并且返回其执行结果。

ZygoteInit 类的 handleSystemServerProcess() 函数用于处理新生成的 System Server 进程剩下的所有工作，代码如下：

```
//第5章/ZygoteInit.java
/**
 * Finish remaining work for the newly forked system server process.
 */
private static Runnable handleSystemServerProcess(
        ZygoteArguments parsedArgs) {
    //set umask to 0077 so new files and directories will default to
    //owner-only permissions
    Os.umask(S_IRWXG | S_IRWXO);

    if (parsedArgs.mNiceName != null) {
        Process.setArgV0(parsedArgs.mNiceName);
    }
```

```java
        final String systemServerClasspath = Os.getenv("SYSTEMSERVERCLASSPATH");
        if (systemServerClasspath != null) {
            if (performSystemServerDexOpt(systemServerClasspath)) {
                //Throw away the cached classloader. If we compiled here, the
                //classloader would not have had AoT-ed artifacts
                //Note: This only works in a very special environment where
                //seLinux enforcement is disabled, e.g., Mac builds
                sCachedSystemServerClassLoader = null;
            }
            //Capturing profiles is only supported for debug or eng builds
            //since seLinux normally prevents it
            boolean profileSystemServer = SystemProperties.getBoolean(
                    "dalvik.vm.profilesystemserver", false);
            if (profileSystemServer && (Build.IS_USERDEBug || Build.IS_ENG)) {
                try {
                    prepareSystemServerProfile(systemServerClasspath);
                } catch (Exception e) {
                    Log.wtf(TAG, "Failed to set up system server profile", e);
                }
            }
        }

        if (parsedArgs.mInvokeWith != null) {
            String[] args = parsedArgs.mRemainingArgs;
            //If we have a non-null system server class path, we'll have to
            //duplicate the existing arguments and append the classpath to
            //it. ART will handle the classpath correctly when we exec a new
            //process
            if (systemServerClasspath != null) {
                String[] amendedArgs = new String[args.length + 2];
                amendedArgs[0] = "-cp";
                amendedArgs[1] = systemServerClasspath;
                System.arraycopy(args, 0, amendedArgs, 2, args.length);
                args = amendedArgs;
            }

            WrapperInit.execApplication(parsedArgs.mInvokeWith,
                    parsedArgs.mNiceName, parsedArgs.mTargetSdkVersion,
                    VMRuntime.getCurrentInstructionSet(), null, args);

            throw new IllegalStateException("Unexpected return from WrapperInit.execApplication");
        } else {
            createSystemServerClassLoader();
            ClassLoader cl = sCachedSystemServerClassLoader;
            if (cl != null) {
```

```
            Thread.currentThread().setContextClassLoader(cl);
        }

        /*
         * Pass the remaining arguments to SystemServer.
         */
        return ZygoteInit.zygoteInit(parsedArgs.mTargetSdkVersion,
                parsedArgs.mRemainingArgs, cl);
    }

    /* should never reach here */
}
```

本函数在为 System Server 准备 profiles 之后，又为 System Server 创建了 classloader，之后调用了 ZygoteInit.zygoteInit()函数，将剩余的参数传递给了 System Server。

向上回溯，ZygoteInit 的 forkSystemServer()函数的返回值分为 null 和非 null。其中返回 null 的是父进程，返回非 null 的情况下是新进程（子进程），这两种情况在 ZygoteInit 的 main()函数中进行了不同的处理。子进程会运行 ZygoteInit 的 forkSystemServer()函数所返回的 Runnable 类型对象的 run()函数，然后退出；父进程会运行 ZygoteServer 类的 runSelectLoop()函数，进入一个无限循环的状态。在这个无线循环的过程中，会根据所接收的信息不断地去为应用创建新的进程，这部分将在接下来的内容中进行介绍。

5.7.2 应用进程

Zygote 进程 fork 出 System Server 进程之后，本进程还会继续运行，进行 ZygoteServer.runSelectLoop()函数中的循环状态，等待信息为应用程序创建对应的 App 进程。

ZygoteServer 类的 runSelectLoop()函数位于 frameworks/base/core/java/com/android/internal/os/ZygoteServer.java 中，代码如下：

```
//第 5 章/ZygoteServer.java
/**
 * Runs the zygote process's select loop. Accepts new connections as
 * they happen, and reads commands from connections one spawn-request's
 * worth at a time.
 */
Runnable runSelectLoop(String abiList) {
    ArrayList<FileDescriptor> socketFDs = new ArrayList<FileDescriptor>();
    ArrayList<ZygoteConnection> peers = new ArrayList<ZygoteConnection>();

    socketFDs.add(mZygoteSocket.getFileDescriptor());
    peers.add(null);

    while (true) {
```

```
fetchUsapPoolPolicyPropsWithMinInterval();

int[] usapPipeFDs = null;
StructPollfd[] pollFDs = null;

//Allocate enough space for the poll structs, taking into account
//the state of the USAP pool for this Zygote (could be a
//regular Zygote, a WebView Zygote, or an AppZygote)
if (mUsapPoolEnabled) {
    usapPipeFDs = Zygote.getUsapPipeFDs();
    pollFDs = new StructPollfd[socketFDs.size() + 1 + usapPipeFDs.length];
} else {
    pollFDs = new StructPollfd[socketFDs.size()];
}

/*
 * For reasons of correctness the USAP pool pipe and event FDs
 * must be processed before the session and server sockets. This
 * is to ensure that the USAP pool accounting information is
 * accurate when handling other requests like API blacklist
 * exemptions.
 */

int pollIndex = 0;
for (FileDescriptor socketFD : socketFDs) {
    pollFDs[pollIndex] = new StructPollfd();
    pollFDs[pollIndex].fd = socketFD;
    pollFDs[pollIndex].events = (short) POLLIN;
    ++pollIndex;
}

final int usapPoolEventFDIndex = pollIndex;

if (mUsapPoolEnabled) {
    pollFDs[pollIndex] = new StructPollfd();
    pollFDs[pollIndex].fd = mUsapPoolEventFD;
    pollFDs[pollIndex].events = (short) POLLIN;
    ++pollIndex;

    for (int usapPipeFD : usapPipeFDs) {
        FileDescriptor managedFd = new FileDescriptor();
        managedFd.setInt$(usapPipeFD);

        pollFDs[pollIndex] = new StructPollfd();
        pollFDs[pollIndex].fd = managedFd;
        pollFDs[pollIndex].events = (short) POLLIN;
```

```java
            ++pollIndex;
        }
    }

    try {
        Os.poll(pollFDs, -1);
    } catch (ErrnoException ex) {
        throw new RuntimeException("poll failed", ex);
    }

    boolean usapPoolFDRead = false;

    while (--pollIndex >= 0) {
        if ((pollFDs[pollIndex].revents & POLLIN) == 0) {
            continue;
        }

        if (pollIndex == 0) {
            //Zygote server socket

            ZygoteConnection newPeer = acceptCommandPeer(abiList);
            peers.add(newPeer);
            socketFDs.add(newPeer.getFileDescriptor());

        } else if (pollIndex < usapPoolEventFDIndex) {
            //Session socket accepted from the Zygote server socket

            try {
                ZygoteConnection connection = peers.get(pollIndex);
                final Runnable command =
                        connection.processOneCommand(this);

                //TODO (chriswailes): Is this extra check necessary?
                if (mIsForkChild) {
                    //We're in the child. We should always have a
                    //command to run at this stage if processOneCommand
                    //hasn't called "exec"
                    if (command == null) {
                        throw new IllegalStateException(
                                "command == null");
                    }

                    return command;
                } else {
                    //We're in the server - we should never have any
                    //commands to run
```

```java
                    if (command != null) {
                        throw new IllegalStateException(
                                "command != null");
                    }

                    //We don't know whether the remote side of the
                    //socket was closed or not until we attempt to read
                    //from it from processOneCommand. This
                    //shows up as a regular POLLIN event in our regular
                    //processing loop
                    if (connection.isClosedByPeer()) {
                        connection.closeSocket();
                        peers.remove(pollIndex);
                        socketFDs.remove(pollIndex);
                    }
                }
            } catch (Exception e) {
                if (!mIsForkChild) {
                    //We're in the server so any exception here is one
                    //that has taken place pre-fork while processing
                    //commands or reading / writing from the control
                    //socket. Make a loud noise about any such
                    //exceptions so that we know exactly what failed and
                    //why

                    Slog.e(TAG, "Exception executing zygote command: ",
                            e);

                    //Make sure the socket is closed so that the other
                    //end knows immediately that something has gone
                    //wrong and doesn't time out waiting for a response
                    ZygoteConnection conn = peers.remove(pollIndex);
                    conn.closeSocket();

                    socketFDs.remove(pollIndex);
                } else {
                    //We're in the child so any exception caught here
                    //has happened post fork and before we execute
                    //ActivityThread.main (or any other main()
                    //method). Log the details of the exception and
                    //bring down the process
                    Log.e(TAG, "Caught post-fork exception in child
                            process.", e);
                    throw e;
                }
            } finally {
```

```java
                    //Reset the child flag, in the event that the child
                    //process is a child-zygote. The flag will not be
                    //consulted this loop pass after the Runnable is
                    //returned
                    mIsForkChild = false;
                }
            } else {
                //Either the USAP pool event FD or a USAP reporting pipe
                //If this is the event FD the payload will be the number
                //of USAPs removed. If this is a reporting pipe FD the
                //payload will be the PID of the USAP that was just
                //specialized
                long messagePayload = -1;

                try {
                    Byte[] buffer = new
                        Byte[Zygote.USAP_MANAGEMENT_MESSAGE_BYTES];
                    int readBytes = Os.read(pollFDs[pollIndex].fd, buffer,
                                                0, buffer.length);

                    if (readBytes == Zygote.USAP_MANAGEMENT_MESSAGE_BYTES) {
                        DataInputStream inputStream =
                            new DataInputStream(new
                                ByteArrayInputStream(buffer));

                        messagePayload = inputStream.readLong();
                    } else {
                        Log.e(TAG, "Incomplete read from USAP management FD"
                            of size " + readBytes);
                        continue;
                    }
                } catch (Exception ex) {
                    if (pollIndex == usapPoolEventFDIndex) {
                        Log.e(TAG, "Failed to read from USAP pool event FD: "
                                + ex.getMessage());
                    } else {
                        Log.e(TAG, "Failed to read from USAP reporting pipe: "
                                " + ex.getMessage());
                    }

                    continue;
                }

                if (pollIndex > usapPoolEventFDIndex) {
                    Zygote.removeUsapTableEntry((int) messagePayload);
                }

                usapPoolFDRead = true;
            }
```

```
                }

                //Check to see if the USAP pool needs to be refilled
                if (usapPoolFDRead) {
                    int[] sessionSocketRawFDs =
                            socketFDs.subList(1, socketFDs.size()).stream()
                                    .mapToInt(fd -> d.getInt$()).toArray();

                    final Runnable command = fillUsapPool(sessionSocketRawFDs);

                    if (command != null) {
                        return command;
                    }
                }
            }
        }
    }
```

runSelectLoop()函数中有一个无限循环的操作,在代码上体现为 while(true),这个 while(true)后所执行的代码在 while 的循环下不停歇地执行,也就是这个循环保障了 App 不断地运行。这里唯一的退出机制是在 command 不等于空的时候,会返回 command,除此之外,while 循环中的代码将一直执行下去。while 循环中的代码在不断地循环,它会根据 socket 所接收的信息,创建新的应用进程。应用进程的执行,涉及 App 运行的相关内容,在后续 ART 运行的部分将进行详细介绍,在此处不再进行详细介绍,所以 System Server 进程和应用进程都是从 Zygote 进程中 fork 出来的,这两者都是了解 Zygote 进程及整个 ART 所必不可缺少的部分。

本部分内容对 Zygote 进程的基本内容及其所涉及的 System Server 进程和应用进程的相关内容进行了分析,完成了 ART 启动分析的各个模块的介绍。

5.8 小结

本章对 ART 启动过程中所涉及的各项内容进行了分析,从虚拟机的启动、JIT 编译器的启动、线程处理、本地方法注册等方面逐项进行了展开。此后,对于 Zygote 进程的启动进行了介绍,Zygote 作为 Android 的重要进程,它承载着 ART 启动及后续 ART 运行过程中的重要内容。

ART 的启动为后续 ART 的运行准备了所需要的环境和框架,所以本章的内容是后续分析 App 在 ART 上持续不断运行的基础,都是为了后续 ART 运行做准备,所以在学习的过程中,要先搞清楚本章节的内容,然后再对下一部分内容进行学习。

另外,本章在分析 Zygote 进程等内容的时候,涉及了通过 fork 新建进程及进程之间通信的内容,这些相关知识需要从操作系统方面进行学习和补充,在本书中不进行专门的介绍。

第 6 章 ART 的执行

ART 启动完成之后就进入了正常的运行状态。在运行状态下，会根据所要运行的 App 的不同情况，为 App 选取不同的形式去执行。本章将介绍 ART 运行中的几个不同的形式，并对每种情况的代码实现进行分析。

6.1 ART 运行基本流程

根据要执行的 App 的输入格式，以及具体执行过程中的情况，ART 的运行可以有多个流程进行选择，本部分内容将对这几种流程的整体情况进行介绍。

在 ART 上运行的 App 通常会将两种格式的执行文件提交给 ART，分别是.oat 格式和.dex 格式。.oat 格式文件通过 dex2oat 工具将.dex 进行编译所得。这两种格式的文件在 ART 运行的时候是在不同的流程中进行处理的。.oat 格式的文件，因为是提前编译过的 OAT 二进制格式文件，所以交给 ART 之后，会走直接执行的流程，通过运行时直接执行 OAT 中已经编译好的方法，直接执行到 App 结束；.dex 格式的文件交给 ART 之后，因为其不是 OAT 这种已经编译好的二进制文件，而是 dex 格式的中间语言，ART 会将其放到解释器进行解释执行或者交给 JIT 编译器进行即时编译之后执行。上述执行流程的具体流程如图 6.1 所示。

无论 ART 接受的输入是.oat 格式，还是.dex 格式的 App，它们在 ART 上执行的时候都会按照方法逐个去执行。这是 ART 的运行机制，在 App 的方法层面上去逐个运行。.oat 格式文件是已经编译过的，所以，.oat 格式文件中的方法可以直接运行。.dex 格式文件中的方法在执行的时候会面临 JIT 编译器编译过的方法和没编译过的方法两种情况。JIT 编译器编译过的方法，可以从 JIT 代码缓存中取出，然后直接执行；没有编译过的 dex 方法，则需要交给解释器进行解释执行，同时还会根据触发条件判断该方法的代码是否会触发 JIT 编译，如果触发了 JIT 编译条件，就会将该方法进行编译之后放入 JIT 代码缓存中，否则依然进行解释执行。

所以，可以将 ART 的运行分为 3 个流程：第 1 个流程，.oat 格式的文件直接被 ART 执行；第 2 个流程，.dex 格式的文件在 ART 上进行解释执行；第 3 个流程，.dex 格式文件

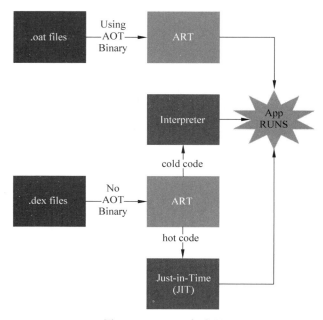

图 6.1　ART 运行流程

(图源：https://source.android.com/devices/tech/dalvik/jit-compiler)

经过 JIT 编译器编译之后，在 ART 的代码缓存中被执行。这 3 个流程在本章的后续部分会分别进行分析，介绍其实现和具体流程。

6.2　Zygote 进程调用应用程序

在 ART 执行的过程中，Zygote 进程会创建新的进程，然后该进程会通过命令去调用应用程序，本部分将对这个过程的代码实现进行分析。

在代码实现上，Zygote 进程会运行 ZygoteServer 类的 runSelectLoop() 函数，在 runSelectLoop() 函数中，会根据 socket 发来的信息，为即将运行的应用创建一个新的进程。这是 ART 开始运行，App 开始运行在 ART 之上的入口。runSelectLoop() 函数中所调用的 ZygoteConnection.processOneCommand() 函数实现了根据 socket 信息新建子进程的功能，processOneCommand() 函数位于 frameworks\base\core\java\com\android\internal\os\ZygoteConnection.java 中，代码如下：

```
//第 6 章/ZygoteConnection.java
/**
 * Reads one start command from the command socket. If successful, a
 * child is forked and a {@code Runnable} that calls the childs main
 * method (or equivalent) is returned in the child process. {@code null}
```

```
 * is always returned in the parent process (the zygote).
 *
 * If the client closes the socket, an {@code EOF} condition is set,
 * which callers can test for by calling {@code
 * ZygoteConnection.isClosedByPeer}.
 */
Runnable processOneCommand(ZygoteServer zygoteServer) {
    String args[];
    ZygoteArguments parsedArgs = null;
    FileDescriptor[] descriptors;

    try {
        args = Zygote.readArgumentList(mSocketReader);

        //TODO (chriswailes): Remove this and add an assert
        descriptors = mSocket.getAncillaryFileDescriptors();
    } catch (IOException ex) {
        throw new IllegalStateException("IOException on command socket",
                ex);
    }

    //readArgumentList returns null only when it has reached EOF with no
    //available data to read. This will only happen when the remote
    //socket has disconnected
    if (args == null) {
        isEof = true;
        return null;
    }

    int pid = -1;
    FileDescriptor childPipeFd = null;
    FileDescriptor serverPipeFd = null;

    parsedArgs = new ZygoteArguments(args);

    if (parsedArgs.mAbiListQuery) {
        handleAbiListQuery();
        return null;
    }

    if (parsedArgs.mPidQuery) {
        handlePidQuery();
        return null;
    }

    if (parsedArgs.mUsapPoolStatusSpecified) {
```

```java
        return handleUsapPoolStatusChange(zygoteServer,
                parsedArgs.mUsapPoolEnabled);
    }

    if (parsedArgs.mPreloadDefault) {
        handlePreload();
        return null;
    }

    if (parsedArgs.mPreloadPackage != null) {
        handlePreloadPackage(parsedArgs.mPreloadPackage,
                parsedArgs.mPreloadPackageLibs,

                parsedArgs.mPreloadPackageLibFileName,
                parsedArgs.mPreloadPackageCacheKey);
        return null;
    }

    if (canPreloadApp() && parsedArgs.mPreloadApp != null) {
        Byte[] rawParcelData =
                Base64.getDecoder().decode(parsedArgs.mPreloadApp);
        Parcel appInfoParcel = Parcel.obtain();
        appInfoParcel.unmarshall(rawParcelData, 0, rawParcelData.length);
        appInfoParcel.setDataPosition(0);
        ApplicationInfo appInfo =
                ApplicationInfo.CREATOR.createFromParcel(appInfoParcel);
        appInfoParcel.recycle();
        if (appInfo != null) {
            handlePreloadApp(appInfo);
        } else {
            throw new IllegalArgumentException("Failed to deserialize --preload-app");
        }
        return null;
    }

    if (parsedArgs.mApiBlacklistExemptions != null) {
        return handleApiBlacklistExemptions(zygoteServer,
                parsedArgs.mApiBlacklistExemptions);
    }

    if (parsedArgs.mHiddenApiAccessLogSampleRate != -1
            || parsedArgs.mHiddenApiAccessStatslogSampleRate != -1) {
        return handleHiddenApiAccessLogSampleRate(zygoteServer,
                parsedArgs.mHiddenApiAccessLogSampleRate,
                parsedArgs.mHiddenApiAccessStatslogSampleRate);
    }
```

```java
        if (parsedArgs.mPermittedCapabilities != 0 ||
                parsedArgs.mEffectiveCapabilities != 0) {
            throw new ZygoteSecurityException("Client may not specify capabilities: "
                    + "permitted = 0x"
                    + Long.toHexString(parsedArgs.mPermittedCapabilities)
                    + ", effective = 0x"
                    + Long.toHexString(parsedArgs.mEffectiveCapabilities));
        }

        Zygote.applyUidSecurityPolicy(parsedArgs, peer);
        Zygote.applyInvokeWithSecurityPolicy(parsedArgs, peer);

        Zygote.applyDebuggerSystemProperty(parsedArgs);
        Zygote.applyInvokeWithSystemProperty(parsedArgs);

        int[][] rlimits = null;

        if (parsedArgs.mRLimits != null) {
            rlimits = parsedArgs.mRLimits.toArray(Zygote.INT_ARRAY_2D);
        }

        int[] fdsToIgnore = null;

        if (parsedArgs.mInvokeWith != null) {
            try {
                FileDescriptor[] pipeFds = Os.pipe2(O_CLOEXEC);
                childPipeFd = pipeFds[1];
                serverPipeFd = pipeFds[0];
                Os.fcntlInt(childPipeFd, F_SETFD, 0);
                fdsToIgnore = new int[]{childPipeFd.getInt$(),
                        serverPipeFd.getInt$()};
            } catch (ErrnoException errnoEx) {
                throw new IllegalStateException("Unable to set up pipe for invoke-with", errnoEx);
            }
        }

        /**
         * In order to avoid leaking descriptors to the Zygote child,
         * the native code must close the two Zygote socket descriptors
         * in the child process before it switches from Zygote-root to
         * the UID and privileges of the application being launched.
         *
         * In order to avoid "bad file descriptor" errors when the
         * two LocalSocket objects are closed, the Posix file
         * descriptors are released via a dup2() call which closes
         * the socket and substitutes an open descriptor to /dev/null.
```

```java
     */

    int [] fdsToClose = { -1, -1 };

    FileDescriptor fd = mSocket.getFileDescriptor();

    if (fd != null) {
        fdsToClose[0] = fd.getInt$();
    }

    fd = zygoteServer.getZygoteSocketFileDescriptor();

    if (fd != null) {
        fdsToClose[1] = fd.getInt$();
    }

    fd = null;

    pid = Zygote.forkAndSpecialize(parsedArgs.mUid, parsedArgs.mGid,
            parsedArgs.mGids, parsedArgs.mRuntimeFlags, rlimits,
            parsedArgs.mMountExternal, parsedArgs.mSeInfo,
            parsedArgs.mNiceName, fdsToClose, fdsToIgnore,
            parsedArgs.mStartChildZygote,
            parsedArgs.mInstructionSet, parsedArgs.mAppDataDir,
            parsedArgs.mTargetSdkVersion);

    try {
        if (pid == 0) {
            //in child
            zygoteServer.setForkChild();

            zygoteServer.closeServerSocket();
            IoUtils.closeQuietly(serverPipeFd);
            serverPipeFd = null;

            return handleChildProc(parsedArgs, descriptors, childPipeFd,
                    parsedArgs.mStartChildZygote);
        } else {
            //In the parent. A pid < 0 indicates a failure and will be
            //handled in handleParentProc
            IoUtils.closeQuietly(childPipeFd);
            childPipeFd = null;
            handleParentProc(pid, descriptors, serverPipeFd);
            return null;
        }
    } finally {
```

```
            IoUtils.closeQuietly(childPipeFd);
            IoUtils.closeQuietly(serverPipeFd);
        }
    }
```

ZygoteConnection.processOneCommand()函数中重要的函数主要有两个：Zygote.forkAndSpecialize()函数和handleChildProc()函数，在下面的内容中将分别对其进行介绍。

6.2.1　Zygote.forkAndSpecialize

在ZygoteConnection.processOneCommand()函数中，创建新进程的操作位于Zygote.forkAndSpecialize()函数的实现中，它会为父进程(Zygote进程)返回新建子进程的pid，并为子进程返回值为0的pid。Zygote.forkAndSpecialize()函数的实现位于rameworks\base\core\java\com\android\internal\os\Zygote.java中，代码如下：

```
//第6章/Zygote.java
/**
 * Forks a new VM instance. The current VM must have been started
 * with the -Xzygote flag. <b>NOTE: new instance keeps all
 * root capabilities. The new process is expected to call capset()</b>.
 *
 * @param uid the UNIX uid that the new process should setuid() to after
 * fork()ing and and before spawning any threads.
 * @param gid the UNIX gid that the new process should setgid() to after
 * fork()ing and and before spawning any threads.
 * @param gids null-ok; a list of UNIX gids that the new process should
 * setgroups() to after fork and before spawning any threads.
 * @param RuntimeFlags bit flags that enable ART features.
 * @param rlimits null-ok an array of rlimit tuples, with the second
 * dimension having a length of 3 and representing
 * (resource, rlim_cur, rlim_max). These are set via the posix
 * setrlimit(2) call.
 * @param seInfo null-ok a string specifying SELinux information for
 * the new process.
 * @param niceName null-ok a string specifying the process name.
 * @param fdsToClose an array of ints, holding one or more POSIX
 * file descriptor numbers that are to be closed by the child
 * (and replaced by /dev/null) after forking. An integer value
 * of -1 in any entry in the array means "ignore this one".
 * @param fdsToIgnore null-ok an array of ints, either null or holding
 * one or more POSIX file descriptor numbers that are to be ignored
 * in the file descriptor table check.
 * @param startChildZygote if true, the new child process will itself be
 * a new zygote process.
```

```
 * @param instructionSet null-ok the instruction set to use.
 * @param appDataDir null-ok the data directory of the app.
 *
 * @return 0 if this is the child, pid of the child
 * if this is the parent, or -1 on error.
 */
public static int forkAndSpecialize(int uid, int gid, int[] gids,
        int RuntimeFlags, int[][] rlimits, int mountExternal,
        String seInfo, String niceName, int[] fdsToClose,
        int[] fdsToIgnore, boolean startChildZygote,
        String instructionSet, String appDataDir, int targetSdkVersion) {
    ZygoteHooks.preFork();
    //Resets nice priority for zygote process
    resetNicePriority();
    int pid = nativeForkAndSpecialize(
            uid, gid, gids, RuntimeFlags, rlimits, mountExternal, seInfo,
            niceName, fdsToClose, fdsToIgnore, startChildZygote,
            instructionSet, appDataDir);
    //Enable tracing as soon as possible for the child process
    if (pid == 0) {
        Zygote.disableExecuteOnly(targetSdkVersion);
        Trace.setTracingEnabled(true, RuntimeFlags);

        //Note that this event ends at the end of handleChildProc
        Trace.traceBegin(Trace.TRACE_TAG_ACTIVITY_MANAGER, "PostFork");
    }
    ZygoteHooks.postForkCommon();
    return pid;
}
```

在 Zygote.forkAndSpecialize() 函数中创建新进程的工作是由 Zygote.nativeForkAndSpecialize() 函数实现的，它是一个 JNI 函数，它的底层实现是 com_android_internal_os_Zygote_nativeForkAndSpecialize() 函数。com_android_internal_os_Zygote_nativeForkAndSpecialize() 函数的实现位于 frameworks/base/core/jni/com_android_internal_os_Zygote.cpp 中，代码如下：

```
//第 6 章/com_android_internal_os_Zygote.cpp
static jint com_android_internal_os_Zygote_nativeForkAndSpecialize(
    JNIEnv* env, jclass, jint uid, jint gid, jintArray gids,
    jint Runtime_flags, jobjectArray rlimits,
    jint mount_external, jstring se_info, jstring nice_name,
    jintArray managed_fds_to_close, jintArray managed_fds_to_ignore,
    jboolean is_child_zygote, jstring instruction_set,
    jstring app_data_dir) {
    jlong capabilities = CalculateCapabilities(env, uid, gid, gids,
```

```cpp
                                                      is_child_zygote);
    if (UNLIKELY(managed_fds_to_close == nullptr)) {
        ZygoteFailure(env, "zygote", nice_name,
                      "Zygote received a null fds_to_close vector.");
    }

    std::vector<int> fds_to_close =
            ExtractJIntArray(env, "zygote", nice_name, managed_fds_to_close).value();
    std::vector<int> fds_to_ignore =
            ExtractJIntArray(env, "zygote", nice_name, managed_fds_to_ignore).value_or
(std::vector<int>());

    std::vector<int> usap_pipes = MakeUsapPipeReadFDVector();

    fds_to_close.insert(fds_to_close.end(), usap_pipes.begin(),
                        usap_pipes.end());
    fds_to_ignore.insert(fds_to_ignore.end(), usap_pipes.begin(),
                         usap_pipes.end());

    fds_to_close.push_back(gUsapPoolSocketFD);

    if (gUsapPoolEventFD != -1) {
      fds_to_close.push_back(gUsapPoolEventFD);
      fds_to_ignore.push_back(gUsapPoolEventFD);
    }

    pid_t pid = ForkCommon(env, false, fds_to_close, fds_to_ignore);

    if (pid == 0) {
      SpecializeCommon(env, uid, gid, gids, Runtime_flags, rlimits,
                       capabilities, capabilities,
                       mount_external, se_info, nice_name, false,
                       is_child_zygote == JNI_TRUE, instruction_set, app_data_dir);
    }
    return pid;
}
```

这里 ForkCommon()函数和 SpecializeCommon()函数的实现都位于 frameworks/base/core/jni/com_android_internal_os_Zygote.cpp 中。在 ForkCommon()函数中进行新建子进程的动作，并且新创建的子进程接下来会执行 SpecializeCommon()函数。在 SpecializeCommon()函数中会去调用 CallStaticVoidMethod()函数，该函数最终会调用 Zygote 的 callPostForkChildHooks()函数。

6.2.2　ZygoteConnection.handleChildProc

回到本节最初的 ZygoteConnection.processOneCommand()函数中，在新建完了子进

程之后，会让新建的子进程去执行 App 程序，这个操作位于 ZygoteConnection.handleChildProc() 函数中。ZygoteConnection.handleChildProc() 函数的实现位于 frameworks/base/core/java/com/android/internal/os/ZygoteConnection.java 中，代码如下：

```java
//第 6 章/ZygoteConnection.java
  /**
   * Handles post-fork setup of child proc, closing sockets as
   * appropriate, reopen stdio as appropriate, and ultimately throwing
   * MethodAndArgsCaller if successful or returning if failed.
   *
   * @param parsedArgs non-null; zygote args
   * @param descriptors null-ok; new file descriptors for stdio if
   * available
   * @param pipeFd null-ok; pipe for communication back to Zygote.
   * @param isZygote whether this new child process is itself a new
   * Zygote
   */
  private Runnable handleChildProc(ZygoteArguments parsedArgs,
          FileDescriptor[] descriptors,
          FileDescriptor pipeFd, boolean isZygote) {
      /**
       * By the time we get here, the native code has closed the two
       * actual Zygote socket connections, and substituted /dev/null in
       * their place. The LocalSocket objects still need to be closed
       * properly.
       */

      closeSocket();
      if (descriptors != null) {
          try {
              Os.dup2(descriptors[0], STDIN_FILENO);
              Os.dup2(descriptors[1], STDOUT_FILENO);
              Os.dup2(descriptors[2], STDERR_FILENO);

              for (FileDescriptor fd: descriptors) {
                  IoUtils.closeQuietly(fd);
              }
          } catch (ErrnoException ex) {
              Log.e(TAG, "Error reopening stdio", ex);
          }
      }

      if (parsedArgs.mNiceName != null) {
          Process.setArgV0(parsedArgs.mNiceName);
```

```
            }
            //End of the postFork event
            Trace.traceEnd(Trace.TRACE_TAG_ACTIVITY_MANAGER);
            if (parsedArgs.mInvokeWith != null) {
                WrapperInit.execApplication(parsedArgs.mInvokeWith,
                        parsedArgs.mNiceName, parsedArgs.mTargetSdkVersion,
                        VMRuntime.getCurrentInstructionSet(),
                        pipeFd, parsedArgs.mRemainingArgs);

                //Should not get here
                throw new IllegalStateException(
                    "WrapperInit.execApplication unexpectedly returned");
            } else {
                if (!isZygote) {
                    return ZygoteInit.zygoteInit(parsedArgs.mTargetSdkVersion,
                            parsedArgs.mRemainingArgs, null /* classLoader */);
                } else {
                    return ZygoteInit.childZygoteInit(
                        parsedArgs.mTargetSdkVersion,
                        parsedArgs.mRemainingArgs, null /* classLoader */);
                }
            }
        }
```

其中，执行 App 会执行 ZygoteInit.zygoteInit()函数所在程序的分支，调用 ZygoteInit.zygoteInit()函数，并返回其执行结果。ZygoteInit.zygoteInit()函数的实现位于 frameworks/base/core/java/com/android/internal/os/ZygoteInit.java 文件中，它对 RuntimeInit 和 ZygoteInit 进行了一些初始化之后，最终调用了 RuntimeInit.applicationInit()函数，这个函数是 ZygoteInit.zygoteInit()函数的核心。ZygoteInit.zygoteInit()函数的代码如下：

```
//第 6 章/ZygoteInit.java
    /**
     * The main function called when started through the zygote process.
     * This could be unified with main(), if the native code in
     * nativeFinishInit() were rationalized with Zygote startup.<p>
     *
     * Current recognized args:
     * <ul>
     * <li> <code> [ -- ] &lt;start class name&gt; &lt;args&gt;
     * </ul>
     *
     * @param targetSdkVersion target SDK version
     * @param argv arg strings
```

```
     */
    public static final Runnable zygoteInit(int targetSdkVersion,
        String[] argv, ClassLoader classLoader) {
        if (RuntimeInit.DEBug) {
            Slog.d(RuntimeInit.TAG,
                "RuntimeInit: Starting application from zygote");
        }

        Trace.traceBegin(Trace.TRACE_TAG_ACTIVITY_MANAGER, "ZygoteInit");
        RuntimeInit.redirectLogStreams();

        RuntimeInit.commonInit();
        ZygoteInit.nativeZygoteInit();
        return RuntimeInit.applicationInit(targetSdkVersion, argv, classLoader);
    }
```

RuntimeInit.applicationInit()函数则通过准备相关参数之后,去调用RuntimeInit类的findStaticMain()函数。RuntimeInit.applicationInit()函数的实现位于frameworks/base/core/java/com/android/internal/os/RuntimeInit.java中,代码如下:

```
//第6章/RuntimeInit.java
    protected static Runnable applicationInit(int targetSdkVersion,
        String[] argv, ClassLoader classLoader) {
        //If the application calls System.exit(), terminate the process
        //immediately without running any shutdown hooks. It is not
        //possible to shutdown an Android application gracefully. Among
        //other things, the Android Runtime shutdown hooks close the
        //Binder driver, which can cause leftover running threads to
        //crash before the process actually exits
        nativeSetExitWithoutCleanup(true);

        //We want to be fairly aggressive about heap utilization, to
        //avoid holding on to a lot of memory that isn't needed
        VMRuntime.getRuntime().setTargetHeapUtilization(0.75f);
        VMRuntime.getRuntime().setTargetSdkVersion(targetSdkVersion);

        final Arguments args = new Arguments(argv);

        //The end of of the RuntimeInit event (see #zygoteInit)
        Trace.traceEnd(Trace.TRACE_TAG_ACTIVITY_MANAGER);

        //Remaining arguments are passed to the start class's static main
        return findStaticMain(args.startClass, args.startArgs,
                        classLoader);
    }
```

findStaticMain()函数也位于 RuntimeInit.java 中，它会根据参数所传入的类名去找到指定类，然后找到该类的 main()函数 m，最终为 m 构建一个 MethodAndArgsCaller 的新对象。findStaticMain()函数的代码如下：

```java
//第6章/RuntimeInit.java
    /**
     * Invokes a static "main(argv[]) method on class "className".
     * Converts various failing exceptions into RuntimeExceptions,
     * with the assumption that they will then cause the VM instance
     * to exit.
     *
     * @param className Fully-qualified class name
     * @param argv Argument vector for main()
     * @param classLoader the classLoader to load {@className} with
     */
    protected static Runnable findStaticMain(String className,
        String[] argv, ClassLoader classLoader) {
        Class<?> cl;

        try {
            cl = Class.forName(className, true, classLoader);
        } catch (ClassNotFoundException ex) {
            throw new RuntimeException(
                    "Missing class when invoking static main "
                    + className, ex);
        }

        Method m;
        try {
            m = cl.getMethod("main", new Class[] { String[].class });
        } catch (NoSuchMethodException ex) {
            throw new RuntimeException(
                    "Missing static main on " + className, ex);
        } catch (SecurityException ex) {
            throw new RuntimeException(
                    "Problem getting static main on " + className, ex);
        }

        int modifiers = m.getModifiers();
        if (! (Modifier.isStatic(modifiers) &&
              Modifier.isPublic(modifiers))) {
            throw new RuntimeException(
                "Main method is not public and static on " +
                className);
        }
```

```
/*
 * This throw gets caught in ZygoteInit.main(), which responds
 * by invoking the exception's run() method. This arrangement
 * clears up all the stack frames that were required in setting
 * up the process.
 */
return new MethodAndArgsCaller(m, argv);
}
```

MethodAndArgsCaller 类的实现同样位于 RuntimeInit.java 中,它是对 Runnable 的实现,代码如下:

```
//第6章/RuntimeInit.java
/**
 * Helper class which holds a method and arguments and can call
 * them. This is used as part of a trampoline to get rid of the
 * initial process setup stack frames.
 */
static class MethodAndArgsCaller implements Runnable {
    /** method to call */
    private final Method mMethod;

    /** argument array */
    private final String[] mArgs;

    public MethodAndArgsCaller(Method method, String[] args) {
        mMethod = method;
        mArgs = args;
    }

    public void run() {
        try {
            mMethod.invoke(null, new Object[] { mArgs });
        } catch (IllegalAccessException ex) {
            throw new RuntimeException(ex);
        } catch (InvocationTargetException ex) {
            Throwable cause = ex.getCause();
            if (cause instanceof RuntimeException) {
                throw (RuntimeException) cause;
            } else if (cause instanceof Error) {
                throw (Error) cause;
            }
            throw new RuntimeException(ex);
        }
    }
}
```

findStaticMain()函数新建了一个 MethodAndArgsCaller 对象,并将其返回。因为 MethodAndArgsCaller 是 Runnable 的实现,所以 findStaticMain()函数最终返回的是 Runnable 类型的对象。这个返回的对象,根据代码调用关系不断向上返回,最终会回到 ZygoteInit 的 main()函数中,调用 Runnable 类型对象的 Run()函数,所以应用进程会调用 MethodAndArgsCaller 对象的 Run()函数,这个函数会调用 java.lang.reflect.Method 类型的 mMethod 的 invoke()函数。

java.lang.reflect.Method 的 invoke()函数属于 libcore 自带的函数,在 Android 系统中,它位于 libcore/ojluni/src/main/java/java/reflect/Method.java 中,但是实际上的实现最终是在 art/Runtime/native/java_lang_reflect_Method.cc 中,对应着 Method_invoke()函数,而这里的 mMethod 就是 ActivityThread main 的 main()函数,会一直调用直到 Activity 的 onCreate()函数,然后开始进入 App 的执行。

本部分介绍了 Zygote 进程调用应用程序的部分过程,这是 App 运行在 ART 之上的必经之路,理解了这些内容对于理解 App 是如何运行在 ART 之上非常有帮助。这个过程中的有些细节问题会在后续的部分进行介绍。

6.3 类的查找与定义

在 ART 调用应用程序的过程中,会涉及类的查找与链接,这个过程在运行应用程序的准备过程中十分重要,本部分内容将对这个过程进行分析。

上文提到过 frameworks/base/core/java/com/android/internal/os/RuntimeInit.java 中的 findStaticMain()函数,它会根据类的名字去查找类,代码如下:

```
//第 6 章/RuntimeInit.java
    /**
     * Invokes a static "main(argv[])" method on class "className".
     * Converts various failing exceptions into RuntimeExceptions,
     * with the assumption that they will then cause the VM instance
     * to exit.
     *
     * @param className Fully-qualified class name
     * @param argv Argument vector for main()
     * @param classLoader the classLoader to load {@className} with
     */
    protected static Runnable findStaticMain(String className,
        String[] argv, ClassLoader classLoader) {
        Class<?> cl;

        try {
            cl = Class.forName(className, true, classLoader);
        } catch (ClassNotFoundException ex) {
```

```
            throw new RuntimeException(
                    "Missing class when invoking static main "
                    + className, ex);
        }

        Method m;
        try {
            m = cl.getMethod("main", new Class[] { String[].class });
        } catch (NoSuchMethodException ex) {
            throw new RuntimeException(
                    "Missing static main on " + className, ex);
        } catch (SecurityException ex) {
            throw new RuntimeException(
                    "Problem getting static main on " + className, ex);
        }

        int modifiers = m.getModifiers();
        if (! (Modifier.isStatic(modifiers) &&
              Modifier.isPublic(modifiers))) {
            throw new RuntimeException(
                    "Main method is not public and static on " +
                    className);
        }

        /*
         * This throw gets caught in ZygoteInit.main(), which responds
         * by invoking the exception's run() method. This arrangement
         * clears up all the stack frames that were required in setting
         * up the process.
         */
        return new MethodAndArgsCaller(m, argv);
    }
```

根据类名查找类的过程是在 Class.forName()函数中实现的。Class 是一个基本类,它的实现位于 libcore/ojluni/src/main/java/java/lang/Class.java 中,Class.forName()函数的实现也位于这个文件中,代码如下:

```
//第6章/Class.java
    /**
     * Returns the {@code Class} object associated with the class or
     * interface with the given string name, using the given class
       * loader.
     * Given the fully qualified name for a class or interface (in the
     * same format returned by {@code getName}) this method attempts
     * to locate, load, and link the class or interface. The
```

```
 * specified class loader is used to load the class or interface.
 * If the parameter {@code loader} is null, the class is loaded
 * through the bootstrap class loader. The class is initialized
 * only if the {@code initialize} parameter is {@code true} and if
 * it has not been initialized earlier.
 * <p> If {@code name} denotes a primitive type or void, an
 * attempt will be made to locate a user-defined class in the
 * unnamed package whose name is {@code name}. Therefore, this
 * method cannot be used to obtain any of the {@code Class}
 * objects representing primitive types or void.
 *
 * <p> If {@code name} denotes an array class, the component type
 * of the array class is loaded but not initialized.
 *
 * <p> For example, in an instance method the expression:
 *
 * <blockquote>
 * {@code Class.forName("Foo")}
 * </blockquote>
 *
 * is equivalent to:
 *
 * <blockquote>
 * {@code Class.forName("Foo", true,
 *                  this.getClass().getClassLoader())}
 * </blockquote>
 *
 * Note that this method throws errors related to loading, linking
 * or initializing as specified in Sections 12.2, 12.3 and 12.4 of
 * <em>The * Java Language Specification</em>.
 * Note that this method does not check whether the requested
 * class is accessible to its caller.
 *
 * <p> If the {@code loader} is {@code null}, and a security
 * manager is present, and the caller's class loader is not null,
 * then this method calls the security manager's {@code
 * checkPermission} method with a {@code
 * RuntimePermission("getClassLoader")} permission to
 * ensure it's ok to access the bootstrap class loader.
 *
 * @param name     fully qualified name of the desired class
 * @param initialize if {@code true} the class will be
 *                  initialized. See Section 12.4 of <em>The Java Language
 *                  Specification</em>.
 * @param loader    class loader from which the class must be
 * loaded
```

```
 * @return          class object representing the desired class
 *
 * @exception LinkageError if the linkage fails
 * @exception ExceptionInInitializerError if the initialization
 * provoked by this method fails
 * @exception ClassNotFoundException if the class cannot be
 * located by the specified class loader
 *
 * @see      java.lang.Class#forName(String)
 * @see      java.lang.ClassLoader
 * @since    1.2
 */
@CallerSensitive
public static Class<?> forName(String name, boolean initialize,
                               ClassLoader loader)
    throws ClassNotFoundException
{
    if (loader == null) {
        loader = BootClassLoader.getInstance();
    }
    Class<?> result;
    try {
        result = classForName(name, initialize, loader);
    } catch (ClassNotFoundException e) {
        Throwable cause = e.getCause();
        if (cause instanceof LinkageError) {
            throw (LinkageError) cause;
        }
        throw e;
    }
    return result;
}
```

Class 的 forName()函数调用了它自己的 classForName()函数,以此完成根据名字查找类的功能。classForName()函数是个 native 函数,它的具体实现位于 art/Runtime/native/java_lang_Class.cc 中,代码如下:

```
//第6章/java_lang_Class.cc
    //"name" is in "binary name" format, e.g. "dalvik.system.Debug$1"
    static jclass Class_classForName(JNIEnv* env, jclass,
                                     jstring javaName,
                                     jboolean initialize,
                                     jobject javaLoader) {
  ScopedFastNativeObjectAccess soa(env);
  ScopedUtfChars name(env, javaName);
```

```cpp
  if (name.c_str() == nullptr) {
    return nullptr;
  }

  //We need to validate and convert the name (from x.y.z to x/y/z)
  //This is especially handy for array types, since we want to avoid
  //auto-generating bogus array classes
  if (!IsValidBinaryClassName(name.c_str())) {
    soa.Self()->ThrowNewExceptionF(
        "Ljava/lang/ClassNotFoundException;",
        "Invalid name: %s", name.c_str());
    return nullptr;
  }

  std::string descriptor(DotToDescriptor(name.c_str()));
  StackHandleScope<2> hs(soa.Self());
  Handle<mirror::ClassLoader> class_loader(
      hs.NewHandle(soa.Decode<mirror::ClassLoader>(javaLoader)));
  ClassLinker* class_linker = Runtime::Current()->GetClassLinker();
  Handle<mirror::Class> c(
      hs.NewHandle(class_linker->FindClass(soa.Self(),
        descriptor.c_str(), class_loader)));
  if (c == nullptr) {
    ScopedLocalRef<jthrowable> cause(env, env->ExceptionOccurred());
    env->ExceptionClear();
    jthrowable cnfe = reinterpret_cast<jthrowable>(
        env->NewObject(
            WellKnownClasses::java_lang_ClassNotFoundException,
            WellKnownClasses::java_lang_ClassNotFoundException_init,
            javaName, cause.get()));
    if (cnfe != nullptr) {
      //Make sure allocation didn't fail with an OOME
      env->Throw(cnfe);
    }
    return nullptr;
  }
  if (initialize) {
    class_linker->EnsureInitialized(soa.Self(), c, true, true);
  }
  return soa.AddLocalReference<jclass>(c.Get());
}
```

在 classForName() 函数中查找给定名字的类的操作是通过 ClassLinker 的 FindClass() 函数实现的。FindClass() 函数的实现位于 art/Runtime/class_linker.cc 中，代码如下：

```cpp
//第 6 章/class_linker.cc
    ObjPtr<mirror::Class> ClassLinker::FindClass(Thread* self,
            const char* descriptor,
            Handle<mirror::ClassLoader> class_loader) {
    DCHECK_NE(*descriptor, '\0') << "descriptor is empty string";
    DCHECK(self != nullptr);
    self->AssertNoPendingException();
    self->PoisonObjectPointers();
    if (descriptor[1] == '\0') {
        //only the descriptors of primitive types should be 1 character
        //long, also avoid class lookup for primitive classes that aren't
        //backed by dex files
        return FindPrimitiveClass(descriptor[0]);
    }
    const size_t hash = ComputeModifiedUtf8Hash(descriptor);
    //Find the class in the loaded classes table
    ObjPtr<mirror::Class> klass = LookupClass(self, descriptor, hash,
                                               class_loader.Get());
    if (klass != nullptr) {
        return EnsureResolved(self, descriptor, klass);
    }
    //Class is not yet loaded
    if (descriptor[0] != '[' && class_loader == nullptr) {
    //Non-array class and the boot class loader, search the boot
    //class path
        ClassPathEntry pair = FindInClassPath(descriptor, hash,
                                              boot_class_path_);
        if (pair.second != nullptr) {
            return DefineClass(self,
                                descriptor,
                                hash,
                                ScopedNullHandle<mirror::ClassLoader>(),
                                *pair.first,
                                *pair.second);
        } else {
            //The boot class loader is searched ahead of the application
            //class loader, failures are expected and will be wrapped in a
            //ClassNotFoundException. Use the pre-allocated error to
            //trigger the chaining with a proper stack trace
            ObjPtr<mirror::Throwable> pre_allocated =
                Runtime::Current()->GetPreAllocatedNoClassDefFoundError();
            self->SetException(pre_allocated);
            return nullptr;
        }
    }
    ObjPtr<mirror::Class> result_ptr;
```

```cpp
    bool descriptor_equals;
    if (descriptor[0] == '[') {
      result_ptr = CreateArrayClass(self, descriptor, hash,
                                    class_loader);
      DCHECK_EQ(result_ptr == nullptr, self->IsExceptionPending());
      DCHECK(result_ptr == nullptr ||
             result_ptr->DescriptorEquals(descriptor));
      descriptor_equals = true;
    } else {
      ScopedObjectAccessUnchecked soa(self);
      bool known_hierarchy =
        FindClassInBaseDexClassLoader(soa, self, descriptor, hash,
                                      class_loader, &result_ptr);
      if (result_ptr != nullptr) {
        //The chain was understood and we found the class. We still
        //need to add the class to the class table to protect from racy
        //programs that can try and redefine the path list which would
        //change the Class<?> returned for subsequent evaluation of
        //const-class
        DCHECK(known_hierarchy);
        DCHECK(result_ptr->DescriptorEquals(descriptor));
        descriptor_equals = true;
      } else if (!self->IsExceptionPending()) {
        //Either the chain wasn't understood or the class wasn't found
        //If there is a pending exception we didn't clear, it is a not
        //a ClassNotFoundException and we should return it instead of
        //silently clearing and retrying
        //
        //If the chain was understood but we did not find the class
        //let the Java-side rediscover all this and throw the exception
        //with the right stack trace. Note that the Java-side could
        //still succeed for racy programs if another thread is actively
        //modifying the class loader's path list

        //The Runtime is not allowed to call into java from a Runtime-
        //thread so just abort
        if (self->IsRuntimeThread()) {
          //Oops, we can't call into java so we can't run actual class-
          //loader code.
          //This is true for e.g. for the compiler (jit or aot)
          ObjPtr<mirror::Throwable> pre_allocated =
              Runtime::Current()->GetPreAllocatedNoClassDefFoundError();
          self->SetException(pre_allocated);
          return nullptr;
        }
```

```cpp
//Inlined DescriptorToDot(descriptor) with extra validation
//
//Throw NoClassDefFoundError early rather than potentially load
//a class only to fail the DescriptorEquals() check below and
//give a confusing error message. For example, when native code
//erroneously calls JNI GetFieldId() with signature
//"java/lang/String" instead of "Ljava/lang/String;", the
//message below using the "dot" names would be "class loader
//[...] returned class java.lang.String instead of
//java.lang.String"
size_t descriptor_length = strlen(descriptor);
if (UNLIKELY(descriptor[0] != 'L') ||
    UNLIKELY(descriptor[descriptor_length - 1] != ';') ||
    UNLIKELY(memchr(descriptor + 1, '.',
                    descriptor_length - 2) != nullptr)) {
  ThrowNoClassDefFoundError("Invalid descriptor: %s.",
                            descriptor);
  return nullptr;
}

std::string class_name_string(descriptor + 1,
                              descriptor_length - 2);
std::replace(class_name_string.begin(), class_name_string.end(),
             '/', '.');
if (known_hierarchy &&
    fast_class_not_found_exceptions_ &&
    !Runtime::Current()->IsJavaDebuggable()) {
  //For known hierarchy, we know that the class is going to
  //throw an exception. If we aren't debuggable, optimize this
  //path by throwing directly here without going back to Java
  //language. This reduces how many ClassNotFoundExceptions
  //happen
  self->ThrowNewExceptionF("Ljava/lang/ClassNotFoundException;",
                           "%s", class_name_string.c_str());
} else {
  ScopedLocalRef<jobject> class_loader_object(soa.Env(),
        soa.AddLocalReference<jobject>(class_loader.Get()));
  ScopedLocalRef<jobject> result(soa.Env(), nullptr);
  {
    ScopedThreadStateChange tsc(self, kNative);
    ScopedLocalRef<jobject> class_name_object(soa.Env(),
          soa.Env()->NewStringUTF(class_name_string.c_str()));
    if (class_name_object.get() == nullptr) {
      DCHECK(self->IsExceptionPending());
      return nullptr;
    }
```

```cpp
            CHECK(class_loader_object.get() != nullptr);
            result.reset(soa.Env()->CallObjectMethod(
                class_loader_object.get(),
                WellKnownClasses::java_lang_ClassLoader_loadClass,
              class_name_object.get()));
        }
        if (result.get() == nullptr && !self->IsExceptionPending()) {
            //broken loader - throw NPE to be compatible with Dalvik
            ThrowNullPointerException(StringPrintf(
                "ClassLoader.loadClass returned null for %s",
                class_name_string.c_str()).c_str());
            return nullptr;
        }
        result_ptr = soa.Decode<mirror::Class>(result.get());
        //Check the name of the returned class
        descriptor_equals = (result_ptr != nullptr) &&
                             result_ptr->DescriptorEquals(descriptor);
     }
  } else {
    DCHECK(!MatchesDexFileCaughtExceptions(self->GetException(),
                                            this));
  }
}

if (self->IsExceptionPending()) {
    //If the ClassLoader threw or array class allocation failed, pass
    //that exception up. However, to comply with the RI behavior
    //first check if another thread succeeded
    result_ptr = LookupClass(self, descriptor, hash,
                              class_loader.Get());
    if (result_ptr != nullptr && !result_ptr->IsErroneous()) {
       self->ClearException();
       return EnsureResolved(self, descriptor, result_ptr);
    }
    return nullptr;
}

 //Try to insert the class to the class table, checking for
 //mismatch
ObjPtr<mirror::Class> old;
{
   WriterMutexLock mu(self, *Locks::classlinker_classes_lock_);
   ClassTable* const class_table = InsertClassTableForClassLoader(
        class_loader.Get());
   old = class_table->Lookup(descriptor, hash);
   if (old == nullptr) {
```

```cpp
        old = result_ptr;
        if (descriptor_equals) {
          class_table->InsertWithHash(result_ptr, hash);
          WriteBarrier::ForEveryFieldWrite(class_loader.Get());
        } //else throw below, after releasing the lock
      }
    }
    if (UNLIKELY(old != result_ptr)) {
      //Return `old` (even if `!descriptor_equals`) to mimic the RI
      //behavior for parallel capable class loaders.(All class
      //loaders are considered parallel capable on Android.)
      ObjPtr<mirror::Class> loader_class = class_loader->GetClass();
      const char* loader_class_name = loader_class->GetDexFile()
          .StringByTypeIdx(loader_class->GetDexTypeIndex());
      LOG(WARNING) << "Initiating class loader of type "
              << DescriptorToDot(loader_class_name)
              << " is not well-behaved; it returned a different
                 Class for racing loadClass(\""
              << DescriptorToDot(descriptor) << "\").";
      return EnsureResolved(self, descriptor, old);
    }
    if (UNLIKELY(!descriptor_equals)) {
      std::string result_storage;
      const char* result_name =
          result_ptr->GetDescriptor(&result_storage);
      std::string loader_storage;
      const char* loader_class_name =
          class_loader->GetClass()->GetDescriptor(&loader_storage);
      ThrowNoClassDefFoundError(
          "Initiating class loader of type %s returned class %s instead
              of %s.",
          DescriptorToDot(loader_class_name).c_str(),
          DescriptorToDot(result_name).c_str(),
          DescriptorToDot(descriptor).c_str());
      return nullptr;
    }
    //Success
    return result_ptr;
  }
```

FindClass()函数的主要功能是查找已经加载的 mirror::Class 类,如果这个类没有加载,则要定义类。这里有两个函数比较重要,一个是 LookupClass() 函数,另一个是 DefineClass()函数。这两个函数同样都位于 art/Runtime/class_linker.cc 中,并且用到了 ObjPtr 模板类来与 mirror::Class 类一起使用,其中 ObjPtr 模板类的实现位于 art/Runtime/obj_ptr.h 和 obj_ptr_inl.h 中。LookupClass()函数用于在已经加载的类表中查

找已经加载的类,具体实现就是在类表 class_table 中去查找类,然后返回查找的结果,代码如下:

```cpp
//第6章/class_linker.cc
ObjPtr<mirror::Class> ClassLinker::LookupClass(Thread* self,
    const char* descriptor, size_t hash, ObjPtr<mirror::ClassLoader> class_loader) {
  ReaderMutexLock mu(self, *Locks::classlinker_classes_lock_);
  ClassTable* const class_table =
      ClassTableForClassLoader(class_loader);
  if (class_table != nullptr) {
    ObjPtr<mirror::Class> result = class_table->Lookup(descriptor,
                                                      hash);
    if (result != nullptr) {
     return result;
    }
  }
  return nullptr;
}
```

DefineClass()函数是通过 dex 文件及 dex 中的类定义等参数来定义一个 mirror::Class 对象,具体通过类的描述(descriptor)、类的加载器(class_loader)、dex_file 和 dex_class_def 等参数进行定义,代码如下:

```cpp
//第6章/class_linker.cc
ObjPtr<mirror::Class> ClassLinker::DefineClass(Thread* self,
    const char* descriptor, size_t hash,
    Handle<mirror::ClassLoader> class_loader, const DexFile& dex_file,
    const dex::ClassDef& dex_class_def) {
  StackHandleScope<3> hs(self);
  auto klass = hs.NewHandle<mirror::Class>(nullptr);

  //Load the class from the dex file
  if (UNLIKELY(!init_done_)) {
    //finish up init of hand crafted class_roots_
    if (strcmp(descriptor, "Ljava/lang/Object;") == 0) {
      klass.Assign(GetClassRoot<mirror::Object>(this));
    } else if (strcmp(descriptor, "Ljava/lang/Class;") == 0) {
      klass.Assign(GetClassRoot<mirror::Class>(this));
    } else if (strcmp(descriptor, "Ljava/lang/String;") == 0) {
      klass.Assign(GetClassRoot<mirror::String>(this));
    } else if (strcmp(descriptor, "Ljava/lang/ref/Reference;") == 0) {
      klass.Assign(GetClassRoot<mirror::Reference>(this));
    } else if (strcmp(descriptor, "Ljava/lang/DexCache;") == 0) {
      klass.Assign(GetClassRoot<mirror::DexCache>(this));
    } else if (strcmp(descriptor, "Ldalvik/system/ClassExt;") == 0) {
```

```cpp
    klass.Assign(GetClassRoot<mirror::ClassExt>(this));
  }
}

//For AOT-compilation of an app, we may use a shortened boot class
//path that excludes some Runtime modules. Prevent definition of
//classes in app class loader that could clash with these modules as
//these classes could be resolved differently during execution
if (class_loader != nullptr &&
    Runtime::Current()->IsAotCompiler() &&
    IsReservedBootClassPathDescriptor(descriptor)) {
  ObjPtr<mirror::Throwable> pre_allocated =
      Runtime::Current()->GetPreAllocatedNoClassDefFoundError();
  self->SetException(pre_allocated);
  return nullptr;
}

//This is to prevent the calls to ClassLoad and ClassPrepare which can
//cause java/user-supplied code to be executed. We put it up here so
//we can avoid all the allocations associated with creating the class
//This can happen with (eg) jit threads
if (!self->CanLoadClasses()) {
  //Make sure we don't try to load anything, potentially causing an
  //infinite loop
  ObjPtr<mirror::Throwable> pre_allocated =
      Runtime::Current()->GetPreAllocatedNoClassDefFoundError();
  self->SetException(pre_allocated);
  return nullptr;
}

if (klass == nullptr) {
  //Allocate a class with the status of not ready
  //Interface object should get the right size here. Regular class
  //will figure out the right size later and be replaced with one of
  //the right size when the class becomes resolved
  klass.Assign(AllocClass(self, SizeOfClassWithoutEmbeddedTables(
                      dex_file, dex_class_def)));
}
if (UNLIKELY(klass == nullptr)) {
  self->AssertPendingOOMException();
  return nullptr;
}
//Get the real dex file. This will return the input if there aren't
//any callbacks or they do nothing
DexFile const* new_dex_file = nullptr;
dex::ClassDef const* new_class_def = nullptr;
```

```cpp
//TODO We should ideally figure out some way to move this after we get
//a lock on the klass so it will only be called once
Runtime::Current()->GetRuntimeCallbacks()->ClassPreDefine(descriptor,
    klass, class_loader, dex_file, dex_class_def, &new_dex_file,
    &new_class_def);
//Check to see if an exception happened during Runtime callbacks
//Return if so
if (self->IsExceptionPending()) {
  return nullptr;
}
ObjPtr<mirror::DexCache> dex_cache = RegisterDexFile(*new_dex_file, class_loader.Get());
if (dex_cache == nullptr) {
  self->AssertPendingException();
  return nullptr;
}
klass->SetDexCache(dex_cache);
SetupClass(*new_dex_file, *new_class_def, klass, class_loader.Get());

//Mark the string class by setting its access flag
if (UNLIKELY(!init_done_)) {
  if (strcmp(descriptor, "Ljava/lang/String;") == 0) {
    klass->SetStringClass();
  }
}

ObjectLock<mirror::Class> lock(self, klass);
klass->SetClinitThreadId(self->GetTid());
//Make sure we have a valid empty iftable even if there are errors
klass->SetIfTable(GetClassRoot<mirror::Object>(this)->GetIfTable());

//Add the newly loaded class to the loaded classes table
ObjPtr<mirror::Class> existing = InsertClass(descriptor, klass.Get(), hash);
if (existing != nullptr) {
  //We failed to insert because we raced with another thread. Calling
  //EnsureResolved may cause this thread to block
  return EnsureResolved(self, descriptor, existing);
}

//Load the fields and other things after we are inserted in the table
//This is so that we don't end up allocating unfree-able linear alloc
//resources and then lose the race condition. The other reason is that
//the field roots are only visited from the class table. So we need to
//be inserted before we allocate / fill in these fields
LoadClass(self, *new_dex_file, *new_class_def, klass);
if (self->IsExceptionPending()) {
  VLOG(class_linker) << self->GetException()->Dump();
```

```cpp
  //An exception occured during load, set status to erroneous while
  //holding klass' lock in case notification is necessary
  if (!klass->IsErroneous()) {
    mirror::Class::SetStatus(klass, ClassStatus::kErrorUnresolved,
                             self);
  }
  return nullptr;
}

//Finish loading (if necessary) by finding parents
CHECK(!klass->IsLoaded());
if (!LoadSuperAndInterfaces(klass, *new_dex_file)) {
  //Loading failed
  if (!klass->IsErroneous()) {
    mirror::Class::SetStatus(klass, ClassStatus::kErrorUnresolved,
                             self);
  }
  return nullptr;
}
CHECK(klass->IsLoaded());

//At this point the class is loaded. Publish a ClassLoad event
//Note: this may be a temporary class. It is a listener's
//responsibility to handle this
Runtime::Current()->GetRuntimeCallbacks()->ClassLoad(klass);

//Link the class (if necessary)
CHECK(!klass->IsResolved());
//TODO: Use fast jobjects?
auto interfaces =
    hs.NewHandle<mirror::ObjectArray<mirror::Class>>(nullptr);

MutableHandle<mirror::Class> h_new_class =
    hs.NewHandle<mirror::Class>(nullptr);
if (!LinkClass(self, descriptor, klass, interfaces, &h_new_class)) {
  //Linking failed
  if (!klass->IsErroneous()) {
    mirror::Class::SetStatus(klass, ClassStatus::kErrorUnresolved,
                             self);
  }
  return nullptr;
}
self->AssertNoPendingException();
CHECK(h_new_class != nullptr) << descriptor;
CHECK(h_new_class->IsResolved() && !h_new_class->IsErroneousResolved())
    << descriptor;
```

```
        //Instrumentation may have updated entrypoints for all methods of all
        //classes. However it could not update methods of this class while we
        //were loading it. Now the class is resolved, we can update
        //entrypoints as required by instrumentation
        if (Runtime::Current()->GetInstrumentation()->AreExitStubsInstalled()) {
          //We must be in the kRunnable state to prevent instrumentation from
          //suspending all threads to update entrypoints while we are doing it
          //for this class
          DCHECK_EQ(self->GetState(), kRunnable);
          Runtime::Current()->GetInstrumentation()
                            ->InstallStubsForClass(h_new_class.Get());
        }

        /*
         * We send CLASS_PREPARE events to the debugger from here. The
         * definition of "preparation" is creating the static fields for a
         * class and initializing them to the standard default values, but not
         * executing any code (that comes later, during "initialization").
         *
         * We did the static preparation in LinkClass.
         *
         * The class has been prepared and resolved but possibly not yet
         * verified at this point.
         */
        Runtime::Current()->GetRuntimeCallbacks()
                           ->ClassPrepare(klass, h_new_class);

        //Notify native debugger of the new class and its layout
        jit::Jit::NewTypeLoadedIfUsingJit(h_new_class.Get());

        return h_new_class.Get();
      }
```

在DefineClass()函数中,从Dex文件转化为一个mirror::Class对象的过程中,SetupClass()函数、InsertClass()函数、LoadClass()函数和LinkClass()函数等都承担了重要的工作,是不可缺少的环节。SetupClass()函数主要用于从Dex文件中为mirror::Class对象构建基本的信息;InsertClass()函数则用于将mirror::Class对象插入类表中;LoadClass()函数在插入mirror::Class对象之后再加载类的成员变量和成员函数等信息;LinkClass()函数用于连接父类、成员变量、成员函数等。这几个函数的实现也都位于art/Runtime/class_linker.cc文件中,可以根据实际的需要进一步分析。

本部分内容对在ART运行调用App过程中所涉及的类的查找与定义这个过程进行了分析。其中,ClassLinker::FindClass()函数是关键函数,这是这一过程中的关键节点,这个函数直接负责获取mirror::Class对象的所有操作。同时,这个函数在ART启动的时候也被调用了。在 ART 启动过程中并未对这个函数进行分析,在这里进行了分析,两边的过程都是类似的。

6.4 方法的加载和链接

上文介绍了应用中类的查找与定义,在其中类的加载部分,涉及了方法的加载与链接,本部分将对方法的加载与链接进行分析。

ClassLinker::LoadClass()函数对类(Handle<mirror::Class>)进行了加载,在类的加载过程中涉及了对类中所包含方法的加载与链接,代码如下:

```
//第 6 章/class_linker.cc
void ClassLinker::LoadClass(Thread* self,
                    const DexFile& dex_file,
                    const dex::ClassDef& dex_class_def,
                    Handle<mirror::Class> klass) {
...
    }, [&](const ClassAccessor::Method& method)
REQUIRES_SHARED(Locks::mutator_lock_) {
        ArtMethod* art_method =
klass->GetDirectMethodUnchecked(class_def_method_index,
                                     image_pointer_size_);
        LoadMethod(dex_file, method, klass, art_method);
        LinkCode(this, art_method, oat_class_ptr,
                class_def_method_index);
        uint32_t it_method_index = method.GetIndex();
        if (last_dex_method_index == it_method_index) {
          //duplicate case
          art_method->SetMethodIndex(last_class_def_method_index);
        } else {
          art_method->SetMethodIndex(class_def_method_index);
          last_dex_method_index = it_method_index;
          last_class_def_method_index = class_def_method_index;
        }
        ++class_def_method_index;
    }, [&](const ClassAccessor::Method& method) REQUIRES_SHARED(Locks::mutator_lock_) {
        ArtMethod* art_method = klass->GetVirtualMethodUnchecked(
            class_def_method_index - accessor.NumDirectMethods(),
            image_pointer_size_);
        LoadMethod(dex_file, method, klass, art_method);
        LinkCode(this, art_method, oat_class_ptr, class_def_method_index);
        ++class_def_method_index;
    });
...
}
```

这里，有关方法的加载和链接主要在 LoadMethod()函数和 LinkCode()函数中实现。LoadMethod()函数也是 ClassLinker 类的成员函数，它主要为 ArtMethod 方法补全信息，完成其加载。LinkCode()函数的实现位于 art/Runtime/class_linker.cc 中，它用于为 ArtMethod 方法链接对应的 OatMethod，并为 ArtMethod 方法设置函数执行的入口点，代码如下：

```cpp
//第 6 章/class_linker.cc
static void LinkCode(ClassLinker* class_linker,
                     ArtMethod* method,
                     const OatFile::OatClass* oat_class,
                     uint32_t class_def_method_index)
    REQUIRES_SHARED(Locks::mutator_lock_) {
  ScopedAssertNoThreadSuspension sants(__FUNCTION__);
  Runtime* const Runtime = Runtime::Current();
  if (Runtime->IsAotCompiler()) {
    //The following code only applies to a non-compiler Runtime
    return;
  }
  //Method shouldn't have already been linked
  DCHECK(method->GetEntryPointFromQuickCompiledCode() == nullptr);
  if (oat_class != nullptr) {
    //Every kind of method should at least get an invoke stub from the
    //oat_method. non-abstract methods also get their code pointers
    const OatFile::OatMethod oat_method = oat_class->GetOatMethod(class_def_method_index);
    oat_method.LinkMethod(method);
  }

  //Install entry point from interpreter
  const void* quick_code = method->GetEntryPointFromQuickCompiledCode();
  bool enter_interpreter = class_linker->ShouldUseInterpreterEntrypoint(method, quick_code);

  if (!method->IsInvokable()) {
    EnsureThrowsInvocationError(class_linker, method);
    return;
  }

  if (method->IsStatic() && !method->IsConstructor()) {
    //For static methods excluding the class initializer, install the
    //trampoline. It will be replaced by the proper entry point by
    //ClassLinker::FixupStaticTrampolines after initializing class (see
    //ClassLinker::InitializeClass method)
    method->SetEntryPointFromQuickCompiledCode(GetQuickResolutionStub());
  } else if (quick_code == nullptr && method->IsNative()) {
    method->SetEntryPointFromQuickCompiledCode(GetQuickGenericJniStub());
  } else if (enter_interpreter) {
```

```
        //Set entry point from compiled code if there's no code or in
        //interpreter only mode
        method->SetEntryPointFromQuickCompiledCode(
            GetQuickToInterpreterBridge());
    }

    if (method->IsNative()) {
      //Unregistering restores the dlsym lookup stub
      method->UnregisterNative();

      if (enter_interpreter || quick_code == nullptr) {
        //We have a native method here without code. Then it should have
        //either the generic JNI trampoline as entrypoint (non-static), or
        //the resolution trampoline (static)
        //TODO: this doesn't handle all the cases where trampolines may be
        //installed
        const void* entry_point =
            method->GetEntryPointFromQuickCompiledCode();
        DCHECK(class_linker->IsQuickGenericJniStub(entry_point) ||
            class_linker->IsQuickResolutionStub(entry_point));
      }
    }
}
```

在为 ArtMethod 设置入口点的时候,会有几种不同的情况,这几种不同的情况都使用 ArtMethod 的 SetEntryPointFromQuickCompiledCode() 函数设置入口点,只不过传送给 SetEntryPointFromQuickCompiledCode() 函数的参数不一样。当 method 为静态函数的时候,传递给 SetEntryPointFromQuickCompiledCode() 函数的是 GetQuickResolutionStub() 函数的返回值;当 method 为 JNI 的本地函数的时候,传递给 SetEntryPointFromQuickCompiledCode() 函数的是 GetQuickGenericJniStub() 函数的返回值;在 method 为解释器模式的时候,传递给 SetEntryPointFromQuickCompiledCode() 函数的是 GetQuickToInterpreterBridge() 函数的返回值,所以这里会根据 method 的不同特性,为其设置不同的入口点,不同的入口点也就意味着根据方法的不同类型选择了不同的运行模式。

本部分介绍了方法的加载及其代码的链接,这是为后续方法的执行准备好了相关入口点,在后续进行类相关的链接之后就可以进行执行了。

6.5 方法的执行

应用在 ART 的执行,最终可以看作应用的具体方法的执行。本部分将对其过程做一个简要的分析。

目前方法的入口点设置不同,那么后续不同类型的方法执行将根据各自的入口点进行执行，所以无论是 oat 格式，还是解释器模式，都将根据方法的入口点进行执行。本部分将分为.oat 格式方法执行、解释执行和 JIT 编译执行，其中 JIT 编译执行为本部分内容的重点。

.oat 格式文件作为被 dex2oat 编译之后的二进制文件，它其中的方法可以直接被执行，即在 ART 中直接按照方法逐个进行执行。

解释执行的时候输入的 dex2dex 编译器是处理后的 dex 文件。邓凡平在《深入理解 Android Java 虚拟机 ART》中，对解释执行部分做了很好的描述，此处借用此描述：ART 虚拟机支持解释方式执行 dex 字节码。这部分功能从 ART 源码结构的角度来看，它们被封装在一个名为 mterp 的模块中（相关代码位于 Runtime/interpreter/mterp 中）。mterp 是 modular interpreter 的缩写，它在不同的 CPU 平台上会利用对应的汇编指令来编写 dex 字节码的处理。使用汇编来编写代码可大幅提升执行速度。在 dalvik 虚拟机时代，也有 mterp 这个模块，但除了几个主流 CPU 平台上有汇编实现之外，还存在一个用 C++ 实现的代码，其代码很有参考价值，而在 ART 虚拟机的 mterp 模块里，C++ 实现的代码被去掉了，只留下不同 CPU 平台上的汇编实现。mterp_current_ibase 表示当前正在使用的 interpreter 处理的入口地址；mterp_default_ibase 表示默认的 interpreter 处理入口地址；mterp_alt_ibase 表示可追踪执行情况的 interpreter 地址；这 3 个变量都是指向汇编代码中 interpreter 处理的入口函数。代码中根据实际情况会将 mterp_current_ibase 指向其他两个变量。如果设置了跟踪每条指令解释执行的情况，则 mterp_current_ibase 指向 mterp_alt_ibase，否则指向 mterp_default_ibase。

JIT 编译是在运行的时候将原本计划用于解释执行的代码编译成二进制代码进行执行。目前常见的 JIT 编译主要有基于方法的 JIT、基于踪迹的 JIT 和基于区域的 JIT 等。当前 ART 的 JIT 编译所采用的是基于方法的 JIT。

基于方法的 JIT 是以方法作为一个编译单元进行 JIT 编译的，很多经典的 JIT 编译是基于方法的，这是因为方法作为一个编译单元有着较高的独立性。

在 ART 中，JIT 编译所涉及的流程主要有两个。第 1 个是在程序执行的过程中遇到已经使用 JIT 编译器编译过的方法，可以直接执行。第 2 个是在遇到未编译过的方法时，需要判定是否为热代码，如果为热代码，则进行 JIT 编译，然后放到 JIT 代码缓存里（JIT Code Cache），为接下来的执行做好准备。JIT 编译的过程所真正能体现的地方是在将热代码进行 JIT 编译这个过程中，也就是上面提到的第 2 个流程中，这也是本部分内容将重点关注的地方，具体如图 6.2 所示。

从代码实现层面进行分析，解释器的实现代码中有一个重要的函数，即 Execute() 函数，它的实现位于 art/Runtime/interpreter/interpreter.cc 中，主要负责方法的执行，这里面涉及了 JIT 编译、执行和解释执行，所以在解释器中进行 JIT 编译和执行的过程也体现在这个函数中，代码如下：

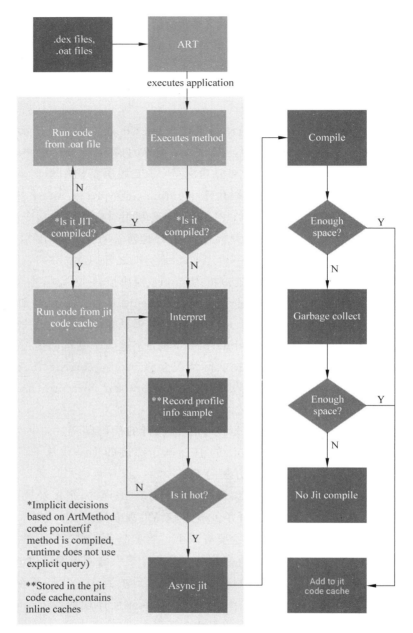

图 6.2 ART 执行示意图

(图源:https://source.android.com/devices/tech/dalvik/jit-compiler)

```
//第6章/interpreter.cc
static inline JValue Execute(
    Thread* self,
```

```
      const CodeItemDataAccessor& accessor,
      ShadowFrame& shadow_frame,
      JValue result_register,
      bool stay_in_interpreter = false,
      bool from_deoptimize = false) REQUIRES_SHARED(Locks::mutator_lock_) {
  DCHECK(!shadow_frame.GetMethod()->IsAbstract());
  DCHECK(!shadow_frame.GetMethod()->IsNative());

  //Check that we are using the right interpreter
  if (kIsDebugBuild && self->UseMterp() != CanUseMterp()) {
    //The flag might be currently being updated on all threads. Retry
    //with lock
    MutexLock tll_mu(self, *Locks::thread_list_lock_);
    DCHECK_EQ(self->UseMterp(), CanUseMterp());
  }

  if (LIKELY(!from_deoptimize)) {
   //Entering the method, but not via deoptimization
...
      if (!stay_in_interpreter && !self->IsForceInterpreter()) {
        jit::Jit* jit = Runtime::Current()->GetJit();
        if (jit != nullptr) {
          jit->MethodEntered(self, shadow_frame.GetMethod());
          if (jit->CanInvokeCompiledCode(method)) {
            JValue result;

            //Pop the shadow frame before calling into compiled code
            self->PopShadowFrame();
            //Calculate the offset of the first input reg. The input
            //registers are in the high regs. It's ok to access the code
            //item here since JIT code will have been touched by the
            //interpreter and compiler already
            uint16_t arg_offset = accessor.RegistersSize() -
                                  accessor.InsSize();
            ArtInterpreterToCompiledCodeBridge(self, nullptr, &shadow_frame,
                                               arg_offset, &result);
            //Push the shadow frame back as the caller will expect it
            self->PushShadowFrame(&shadow_frame);

            return result;
          }
        }
      }
    }
  }
...
}
```

这里所列出的 Execute()函数的代码只选择了其中与 JIT 编译有关的部分代码,对其他部分代码进行了省略。在这里面,最重要的函数是 Jit::MethodEntered()函数和 ArtInterpreterToCompiledCodeBridge()函数,前者对方法进行 JIT 编译,后者则是从解释器到已经编译的代码的桥梁,也可以理解为从解释器调用已经由 JIT 编译过的代码。

Jit::MethodEntered()函数的实现位于 art/Runtime/jit/jit.cc 中,它会通过 JitCompileTask 的 Run()函数或者 AddSamples()函数最终调用 Jit::CompileMethod()函数,去实现对于方法的 JIT 编译。Jit::MethodEntered()函数的代码如下:

```
//第6章/jit.cc
void Jit::MethodEntered(Thread* thread, ArtMethod* method) {
  Runtime* Runtime = Runtime::Current();
  if (UNLIKELY(Runtime->UseJitCompilation() &&
               Runtime->GetJit()->JitAtFirstUse())) {
    ArtMethod* np_method =
                  method->GetInterfaceMethodIfProxy(kRuntimePointerSize);
    if (np_method->IsCompilable()) {
      if (!np_method->IsNative()) {
        //The compiler requires a ProfilingInfo object for non-native
        //methods
        ProfilingInfo::Create(thread, np_method, true);
      }
      JitCompileTask compile_task(method,
          JitCompileTask::TaskKind::kCompile);
      //Fake being in a Runtime thread so that class-load behavior will
      //be the same as normal jit
      ScopedSetRuntimeThread ssrt(thread);
      compile_task.Run(thread);
    }
    return;
  }

  ProfilingInfo* profiling_info =
      method->GetProfilingInfo(kRuntimePointerSize);
  //Update the entrypoint if the ProfilingInfo has one. The interpreter
  //will call it instead of interpreting the method. We don't update it
  //for instrumentation as the entrypoint must remain the
  //instrumentation entrypoint
  if ((profiling_info != nullptr) &&
      (profiling_info->GetSavedEntryPoint() != nullptr) &&
      (method->GetEntryPointFromQuickCompiledCode() != GetQuickInstrumentationEntryPoint())) {
    Runtime::Current()->GetInstrumentation()->UpdateMethodsCode(
        method, profiling_info->GetSavedEntryPoint());
  } else {
    AddSamples(thread, method, 1, /* with_backedges= */false);
  }
}
```

Jit::CompileMethod()函数是 JIT 编译方法的一个入口函数,它通过调用 JIT 编译器对方法进行编译,代码如下:

```
//第6章/jit.cc
bool Jit::CompileMethod(ArtMethod* method, Thread* self, bool baseline,
                        bool osr) {
  DCHECK(Runtime::Current()->UseJitCompilation());
  DCHECK(!method->IsRuntimeMethod());

  RuntimeCallbacks* cb = Runtime::Current()->GetRuntimeCallbacks();
  //Don't compile the method if it has breakpoints
  if (cb->IsMethodBeingInspected(method)
      && !cb->IsMethodSafeToJit(method)) {
    VLOG(jit) << "JIT not compiling " << method->PrettyMethod()
              << " due to not being safe to jit according to Runtime-callbacks. For example, there"
              << " could be breakpoints in this method.";
    return false;
  }

  //Don't compile the method if we are supposed to be deoptimized
  instrumentation::Instrumentation* instrumentation = Runtime::Current()->GetInstrumentation();
  if (instrumentation->AreAllMethodsDeoptimized() || instrumentation->IsDeoptimized(method)) {
    VLOG(jit) << "JIT not compiling " << method->PrettyMethod()
              << " due to deoptimization";
    return false;
  }

  //If we get a request to compile a proxy method, we pass the actual
  //Java method of that proxy method, as the compiler does not expect a
  //proxy method
  ArtMethod* method_to_compile =
    method->GetInterfaceMethodIfProxy(kRuntimePointerSize);
  if (!code_cache_->NotifyCompilationOf(method_to_compile, self, osr)) {
    return false;
  }

  VLOG(jit) << "Compiling method "
            << ArtMethod::PrettyMethod(method_to_compile)
            << " osr=" << std::boolalpha << osr;
  bool success = jit_compile_method_(jit_compiler_handle_, method_to_compile, self, baseline, osr);
  code_cache_->DoneCompiling(method_to_compile, self, osr);
  if (!success) {
    VLOG(jit) << "Failed to compile method "
              << ArtMethod::PrettyMethod(method_to_compile)
              << " osr=" << std::boolalpha << osr;
  }
```

```
    if (kIsDebugBuild) {
      if (self->IsExceptionPending()) {
        mirror::Throwable* exception = self->GetException();
        LOG(FATAL) << "No pending exception expected after compiling "
                   << ArtMethod::PrettyMethod(method)
                   << ": "
                   << exception->Dump();
      }
    }
    return success;
}
```

Jit::CompileMethod()函数是通过jit_compile_method_()函数进行的JIT编译,而jit_compile_method_()函数绑定的是libart-compiler.so中的jit_compile_method()函数。这个jit_compile_method()函数属于JIT编译器的内容,它的实现位于art/compiler/jit/jit_compiler.cc中,代码如下:

```
//第6章/jit_compiler.cc
extern "C" bool jit_compile_method(
    void* handle, ArtMethod* method, Thread* self,
    bool baseline, bool osr)
    REQUIRES_SHARED(Locks::mutator_lock_) {
  auto* jit_compiler = reinterpret_cast<JitCompiler*>(handle);
  DCHECK(jit_compiler != nullptr);
  return jit_compiler->CompileMethod(self, method, baseline, osr);
}
```

这里是调用了JitCompiler的CompileMethod()函数,也就是调用了JIT编译器的函数来对方法进行JIT编译。JitCompiler的CompileMethod()函数的实现位于art/compiler/jit/jit_compiler.cc中,它内部的JIT编译过程是通过compiler_->JitCompile()实现的JIT编译,剩余的内容都与Log相关,代码如下:

```
//第6章/jit_compiler.cc
bool JitCompiler::CompileMethod(Thread* self, ArtMethod* method,
                                bool baseline, bool osr) {
  SCOPED_TRACE << "JIT compiling " << method->PrettyMethod();

  DCHECK(!method->IsProxyMethod());
  DCHECK(method->GetDeclaringClass()->IsResolved());

  TimingLogger logger("JIT compiler timing logger",
   true, VLOG_IS_ON(jit), TimingLogger::TimingKind::kThreadCpu);
  self->AssertNoPendingException();
```

```
Runtime* Runtime = Runtime::Current();

//Do the compilation
bool success = false;
{
  TimingLogger::ScopedTiming t2("Compiling", &logger);
  JitCodeCache* const code_cache = Runtime->GetJit()->GetCodeCache();
  uint64_t start_ns = NanoTime();
  success = compiler_->JitCompile(self, code_cache, method, baseline,
                                  osr, jit_logger_.get());
  uint64_t duration_ns = NanoTime() - start_ns;
  VLOG(jit) << "Compilation of "
            << method->PrettyMethod()
            << " took "
            << PrettyDuration(duration_ns);
}

//Trim maps to reduce memory usage
//TODO: move this to an idle phase
{
  TimingLogger::ScopedTiming t2("TrimMaps", &logger);
  Runtime->GetJitArenaPool()->TrimMaps();
}

Runtime->GetJit()->AddTimingLogger(logger);
return success;
}
```

通过上面的分析,得知 JIT 编译是在 compiler_—> JitCompile( )函数中完成的。这里的 compiler_是 std::unique_ptr < Compiler >类型,并且在 JitCompiler 的构造函数中进行了重置,具体的代码如下:

```
//第 6 章/jit_compiler.cc
JitCompiler::JitCompiler() {
  compiler_options_.reset(new CompilerOptions());
  ParseCompilerOptions();
  compiler_.reset(
      Compiler::Create(*compiler_options_, /*storage=*/ nullptr,
                       Compiler::kOptimizing));
}
```

从这里可以看出,compiler_ 被重置为一个新建立的 OptimizingCompiler 对象,所以 compiler_—> JitCompile 在实际的执行中执行的是 OptimizingCompiler 的 JitCompile( )函数,而与 OptimizingCompiler 有关的内容在前文已经介绍过了,其 JitCompile( )函数虽然没

有进行介绍，但是 JitCompile() 函数所调用的内容则和前文重合度较高，在此不再进行分析。

本部分内容从方法的执行开始，重点介绍了 JIT 执行，从解释器开始，逐步分析并跟踪至 OptimizingCompiler 类的实现，介绍了从解释器到 JIT 编译器的具体编译方法的过程，这个过程是了解 ART JIT 编译的一部分。此外，仍有很多细节和内容并未介绍，如果读者需要进一步了解，则可以从源码中进一步了解相关内容。

6.6 小结

本章主要针对 ART 运行尤其针对有关 App 运行的时候的一些问题进行了分析，主要涉及 ART 运行的基本流程、应用进程的调用、类的查找、方法的加载和链接、方法的执行等内容，为读者展示了 ART 运行的时候的基本状态。

第 7 章 ART GC 实现

垃圾回收(Garbage Collection,GC)是运行时中的重要功能,承担着程序运行期间内存回收的职责。ART 的 GC 承担着 Android Runtime 的内存回收功能,本部分将对垃圾回收的基本内容、分配器实现、回收器实现及分配器和垃圾回收的触发等内容进行分析和介绍。

7.1 GC 的基本内容

垃圾回收的英文全称是 Garbage Collection,简称 GC,这里所讲的 Garbage 虽然被直接翻译为垃圾,其实指的是内存中程序不再使用的空间。垃圾回收的工作主要就是找到内存空间中的垃圾,然后对垃圾进行回收,以便后续使用。

学界一般将垃圾回收的正式提出归于 John McCarthy 在 1960 年发表的名为 *Recursive functions of symbolic expressions and their computation by machine* 的论文,这个论文在论文的注脚中提出了 Garbage Collection。从 1960 年至今,垃圾回收已经发展了 60 余年,但是有关垃圾回收的策略都是建立在几种基本策略之上的。

垃圾回收的基本策略主要有 3 种,分别是标记-清除回收(Mark Sweep GC)、引用计数回收(Reference Counting GC)和复制回收(Copying GC)。除此之外的其他策略,通常是这 3 种策略的组合或改进。有的文献也将 Mark Compact GC 作为一种基本策略,但是它其实也可以被视为一种基本策略的组合,属于标记-清除回收和复制回收的组合。Mark Compact GC 的中文翻译目前已经有两种,分别是标记-复制回收和标记-压缩回收。

标记-清除回收是 John McCarthy 于 1960 年在论文 *Recursive functions of symbolic expressions and their computation by machine* 中提出的。标记-清除由标记和清除两个阶段组成,标记阶段对所有的活动对象进行标记,清除阶段则清除所有没有标记的活动对象。标记-清除策略的算法简单,但是容易导致堆内的空间碎片化。

引用计数回收是 George E. Collins 于 1960 年在论文 *A method for overlapping and erasure of lists* 中提出的。引用计数是为每个对象分配一个计数器,这个计数器用来表示该对象所被引用的次数,一旦某个对象的引用次数为 0,则表示该对象可以被释放回收。引用计数可以即刻释放不使用的对象,然后进行垃圾回收,其优点为暂停时间短,但是也存在一个致命的缺点,即无法处理循环引用的情况。

复制回收是 Marvin L. Minsky 于 1963 年在论文 *A Lisp garbage collector algorithm using serial secondary storage* 中提出的。复制回收通过将 A 空间内的活动对象复制到 B 空间里,然后将 A 空间中的所有对象都进行回收,之后将 A、B 空间角色互换并继续进行回收。复制回收不会发生碎片化的问题,同时还有优秀的吞吐量并可以实现高速分配,但是存在堆使用效率低下的问题。

上述几个经典的垃圾回收策略都是在 20 世纪 60 年代提出的,在后续的几十年的发展过程中,垃圾回收策略都是基于这几种策略的改进和组合,实际上并没有突破性的进展,但是,新的组合策略和算法实现不断地被学者们提出和应用。

此外,在 GC 的理论讨论和实现的过程中,通常会有一些常用的概念,这里将对堆(heap)、赋值器(mutator)、分配器(allocator)、回收器(collector)等概念进行介绍。

- 堆指的是用于执行程序时存放对象的内存空间,GC 主要针对堆中已经分配的对象进行管理。
- 赋值器主要负责生成对象及改变对象的关系,伴随着赋值器的操作会产生垃圾。在对赋值器进行理解的时候,可以将应用程序视为实体,或者将赋值器认为是在执行应用程序。
- 分配器可以看作赋值器内部的一个概念,它主要负责为新对象分配空间。赋值器在生成对象时会向分配器申请空间,分配器会返回一块内存给赋值器。
- 回收器是真正执行垃圾回收的代码,找到已经不使用的垃圾,将其进行回收,所以回收器是回收策略的真正体现部分,也是回收策略实现的重点。

关于 GC 在这几十年的发展成果,内容较多,有专门的书籍介绍。如果想进一步了解,可以阅读《垃圾回收算法手册——自动内存管理的艺术》和《垃圾回收的算法与实现》这两本书。《垃圾回收算法手册——自动内存管理的艺术》是 Richard Jones、Antony Hosking 和 Eliot Moss 所编写的 *The Garbage Collection Handbook*:*The Art of Automatic Memory Management* 的中译本。Richard Jones 也是 GC 的经典著作 *Garbage Collection*:*Algorithms for Automatic Dynamic Memory Management* 的作者,因此《垃圾回收算法手册——自动内存管理的艺术》涵盖了 *Garbage Collection*:*Algorithms for Automatic Dynamic Memory Management* 的内容。《垃圾回收的算法与实现》是日本学者中村成洋和相川光所编写的,由丁灵翻译成中文,这本书的最大特色是配图很多,并且配图能将算法解释得很清楚,读起来非常容易理解,所以如果之前没有了解过 GC 相关内容的读者,可以从《垃圾回收的算法与实现》这本书入门。当然,《垃圾回收的算法与实现》这本书也有一些缺点,即实现篇所选的具体实现例子已经有些过时了,这也跟这本书成书于 2010 年之前有关。同时,《垃圾回收的算法与实现》虽然易于理解和接受,但是它所覆盖的内容没有《垃圾回收算法手册——自动内存管理的艺术》全面,作者自己也承认存在这个问题。在这几本书之外,则需要根据新发表的论文来跟踪 GC 的进展,读者可以根据自己的需求进行阅读。

7.2 ART GC 回收方案介绍

ART GC 通常会采用多个回收方案,本部分将针对 ART 中 GC 的几个回收方案的相关内容进行理论方面的介绍。

ART GC 常用的回收方案有 CMS、SS、GSS 和 CC。

1. CMS

前文已经介绍过了标记清除(Mark Sweep,MS)方案,它是一种基本的 GC 方案。随着 GC 的不断发展,在标记清除之上产生了很多不同的改进方案,CMS 就是其中一种。CMS 是 Concurrent Mark Sweep 的简称,中文翻译为并发标记清除方案,可以简单地理解为并发的标记清除方案。

为了提高垃圾回收的效率,学者们提出了并发垃圾回收。并发垃圾回收允许赋值器线程和回收器线程同时执行以提高回收效率。在这里需要注意区分并发垃圾回收和并行垃圾回收。与并发垃圾回收不同的是,并行垃圾回收通常使用多个回收器线程进行垃圾回收。CMS 方案允许赋值器和回收器线程同时执行,并且其回收器分为标记和清除两个阶段。

2. SS

SS 是 Semi-space / Mark-Sweep Hybrid 回收方案的简称,它是 Semi-space 和 Mark Sweep 两种垃圾回收的混合方案。其中,Semi-space 是复制回收的一种,它在实现的时候会将堆分为两个大小相等的半区,分别是源空间和目标空间,然后进行复制回收。

3. GSS

GSS 是 SS 回收方案的分代回收版本。分代回收在对象中引入了"年龄"的概念,将对象分为新生代对象和老年代对象。其中,新生代对象指的是刚生成的对象,老年代对象指的是到达一定"年龄"的对象。因为新生代对象大部分会变成垃圾,所以分代回收将新生代对象和老年代对象分别进行 GC,对新生代对象进行的 GC 称为新生代 GC,对老年代对象进行的 GC 称为老年代 GC,一般情况下会提高新生代 GC 的频率,降低老年代 GC 的频率,这样可以提高效率。

分代回收并不能单独作为一种方案而直接使用,它通常会和标记-清除、复制等基本 GC 方案结合起来进行使用。在这里,分代回收会和 SS 回收方案结合起来,形成一个新的方案,即 GSS。

4. CC

CC 是 Concurrent Copying 的简称,中文翻译为并发复制回收,它可以看作复制回收的并发版本。CC 可以在读取屏障的帮助下,通过在不暂停应用线程的情况下并发复制对象来执行堆碎片整理,比非并发版本更加高效。

这几种方案都是在 ART GC 中具体使用的 GC 方案,在后续部分会对这几种方案的具体实现进行分析和介绍。

7.3 ART GC 回收器的实现

前文介绍了 ART GC 的多个不同回收方案,每种回收方案都有其对应的回收器,同时还涉及了默认回收器的设置问题。本部分内容将从默认回收器的设置入手,梳理 ART GC 中回收器实现的整体情况。

7.3.1 回收器的类型

ART GC 默认的回收方式是 CMS,即并发标记清除方案。art/build/art.go 中有可见的默认回收方案的设置代码,代码如下:

```
//第 7 章/art.go
    gcType := envDefault(ctx, "ART_DEFAULT_GC_TYPE", "CMS")

    if envTrue(ctx, "ART_TEST_DEBug_GC") {
        gcType = "SS"
        tlab = true
    }

    cflags = append(cflags, "-DART_DEFAULT_GC_TYPE_IS_" + gcType)
    if tlab {
        cflags = append(cflags, "-DART_USE_TLAB=1")
    }
```

在上述代码中会通过"-DART_DEFAULT_GC_TYPE_IS_" + gcType 生成 ART_DEFAULT_GC_TYPE_IS_CMS 选项,而 ART_DEFAULT_GC_TYPE_IS_CMS 在 art/Runtime/gc/collector_type.h 中进行了解析,代码如下:

```
//第 7 章/collector_type.h
static constexpr CollectorType kCollectorTypeDefault =
#if ART_DEFAULT_GC_TYPE_IS_CMS
    kCollectorTypeCMS
#elif ART_DEFAULT_GC_TYPE_IS_SS
    kCollectorTypeSS
#elif ART_DEFAULT_GC_TYPE_IS_GSS
    kCollectorTypeGSS
#else
    kCollectorTypeCMS
#error "ART default GC type must be set"
#endif
```

从这里可以看到,这里其实已经变成了回收器的类型,这也验证了前文提到的回收方案的实现主要是其回收器的实现。kCollectorTypeDefault 会被解析为 kCollectorTypeCMS、kCollectorTypeSS 或 kCollectorTypeGSS,其中默认的情况下是 kCollectorTypeCMS,而且在 ART_DEFAULT_GC_TYPE_IS_CMS、ART_DEFAULT_GC_TYPE_IS_SS 和 ART_DEFAULT_GC_TYPE_IS_GSS 之外的缺省情况,也会选择 kCollectorTypeCMS。kCollectorTypeCMS 是 CMS 回收方案的回收器类型,kCollectorTypeSS 是 SS 回收方案的回收器类型,kCollectorTypeGSS 是 GSS 回收方案的回收器类型。

此外,kCollectorTypeDefault 是 CollectorType 类型的,CollectorType 类型是一个包含了所有回收器类型的枚举类型,定义也位于 art/gc/collector_type.h 中,代码如下:

```
//第7章/collector_type.h
//Which types of collections are able to be performed
enum CollectorType {
  //No collector selected
  kCollectorTypeNone,
  //Non concurrent mark-sweep
  kCollectorTypeMS,
  //Concurrent mark-sweep
  kCollectorTypeCMS,
  //Semi-space / mark-sweep hybrid, enables compaction
  kCollectorTypeSS,
  //A generational variant of kCollectorTypeSS
  kCollectorTypeGSS,
  //Heap trimming collector, doesn't do any actual collecting
  kCollectorTypeHeapTrim,
  //A (mostly) concurrent copying collector
  kCollectorTypeCC,
  //The background compaction of the concurrent copying collector
  kCollectorTypeCCBackground,
  //Instrumentation critical section fake collector
  kCollectorTypeInstrumentation,
  //Fake collector for adding or removing application image spaces
  kCollectorTypeAddRemoveAppImageSpace,
  //Fake collector used to implement exclusion between GC and debugger
  kCollectorTypeDebugger,
  //A homogeneous space compaction collector used in background
  //transition when both foreground and background collector are CMS
  kCollectorTypeHomogeneousSpaceCompact,
  //Class linker fake collector
  kCollectorTypeClassLinker,
  //JIT Code cache fake collector
  kCollectorTypeJitCodeCache,
  //Hprof fake collector
  kCollectorTypeHprof,
  //Fake collector for installing/removing a system-weak holder
```

```
    kCollectorTypeAddRemoveSystemWeakHolder,
    //Fake collector type for GetObjectsAllocated
    kCollectorTypeGetObjectsAllocated,
    //Fake collector type for ScopedGCCriticalSection
    kCollectorTypeCriticalSection,
};
```

这里将所有的回收器类型都列了出来,但是常用的还是CMS\SS\GSS\CC等几个。在每个回收类型的注释中,也介绍了每种类型的大致情况,其中有不少是fake回收器类型,并没有真实完整的实现。此外,在frameworks/base/core/jni/AndroidRuntime.cpp中,也有与GC相关的设置,并且有两处,分别是"-Xgc:"和"-XX:BackgroundGC=",表示分别设置前台和后台的GC,具体的代码如下:

```
//第7章/AndroidRuntime.cpp
int AndroidRuntime::startVm(JavaVM** pJavaVM, JNIEnv** pEnv, bool zygote)
{
…
    /*
     * Garbage-collection related options.
     */
    parseRuntimeOption("dalvik.vm.gctype", gctypeOptsBuf, "-Xgc:");

    //If it set, honor the "enable_generational_cc" device
    //configuration; otherwise, let the Runtime use its default
    //behavior
    std::string enable_generational_cc =
        server_configurable_flags::GetServerConfigurableFlag(
        RUNTIME_NATIVE_BOOT_NAMESPACE,
        ENABLE_GENERATIONAL_CC,
        /*default_value=*/"");
    if (enable_generational_cc == "true") {
        addOption(kGenerationalCCRuntimeOption);
    } else if (enable_generational_cc == "false") {
        addOption(kNoGenerationalCCRuntimeOption);
    }

    parseRuntimeOption("dalvik.vm.backgroundgctype", backgroundgcOptsBuf,
                       "-XX:BackgroundGC=");
…
}
```

这里的前台GC和后台GC指的是App运行在前台和后台的时候所采用的垃圾回收策略。这里的"-Xgc:"有相关的结构体XGcOption,具体位于art/cmdline/cmdline_types.h中,代码如下:

```
//第7章/cmdline_types.h
struct XGcOption {
  //These defaults are used when the command line arguments for -Xgc:
  //are either omitted completely or partially
  gc::CollectorType collector_type_ = gc::kCollectorTypeDefault;
  bool verify_pre_gc_heap_ = false;
  bool verify_pre_sweeping_heap_ = kIsDebugBuild;
  bool generational_cc = kEnableGenerationalCCByDefault;
  bool verify_post_gc_heap_ = false;
  bool verify_pre_gc_rosalloc_ = kIsDebugBuild;
  bool verify_pre_sweeping_rosalloc_ = false;
  bool verify_post_gc_rosalloc_ = false;
  //Do no measurements for kUseTableLookupReadBarrier to avoid test
  //timeouts. b/31679493
  bool measure_ = kIsDebugBuild && !kUseTableLookupReadBarrier;
  bool gcstress_ = false;
};
```

XGcOption 中 CollectorType 类型的 collector_type 变量也被默认设置为上文提到过的 kCollectorTypeDefault，并且还利用 XGcOption 定义了模板 template <> struct CmdlineType < XGcOption > : CmdlineTypeParser < XGcOption >，这里面还调用了 ParseCollectorType() 函数。这个函数的实现位于 art/cmdline/cmdline_types.h 中，从它的实现可以看到"-Xgc:"最终会选择的回收方案，该函数的代码如下：

```
//第7章/cmdline_types.h
static gc::CollectorType ParseCollectorType(const std::string& option) {
  if (option == "MS" || option == "nonconcurrent") {
    return gc::kCollectorTypeMS;
  } else if (option == "CMS" || option == "concurrent") {
    return gc::kCollectorTypeCMS;
  } else if (option == "SS") {
    return gc::kCollectorTypeSS;
  } else if (option == "GSS") {
    return gc::kCollectorTypeGSS;
  } else if (option == "CC") {
    return gc::kCollectorTypeCC;
  } else {
    return gc::kCollectorTypeNone;
  }
}
```

这里的几个回收器类型比上文提到的 kCollectorTypeDefault 的选项多了 kCollectorTypeCC 和 kCollectorTypeNone。kCollectorTypeCC 代表复制回收，kCollectorTypeNone 代表没有回收方案。

后台回收的实现和前台回收类似,有一个 BackgroundGcOption 结构体,并且有一个基于 BackgroundGcOption 的模板,代码如下:

```
//第 7 章/ cmdline_types.h
template <>
struct CmdlineType < BackgroundGcOption >
  : CmdlineTypeParser < BackgroundGcOption >, private BackgroundGcOption {
  Result Parse(const std::string& substring) {
    //Special handling for HSpaceCompact since this is only valid as a
    //background GC type
    if (substring == "HSpaceCompact") {
      background_collector_type_ = gc::kCollectorTypeHomogeneousSpaceCompact;
    } else {
      gc::CollectorType collector_type = ParseCollectorType(substring);
      if (collector_type != gc::kCollectorTypeNone) {
        background_collector_type_ = collector_type;
      } else {
        return Result::Failure();
      }
    }

    BackgroundGcOption res = * this;
    return Result::Success(res);
  }

  static const char * Name() { return "BackgroundGcOption"; }
};
```

这里可以通过其中的 Parse()函数看出,最终的 background_collector_type_ 可以是 kCollectorTypeHomogeneousSpaceCompact,或者 ParseCollectorType()函数中返回的 kCollector-TypeNone 之外的几种结果,即 kCollectorTypeMS、kCollectorTypeCMS、kCollectorTypeSS、kCollectorTypeGSS 和 kCollectorTypeCC,其他的情况都会报错。

总结前台回收的回收器和后台回收的回收器,后台的选项比前台的选项多了一个 kCollectorTypeHomogeneousSpaceCompact,这是后台回收专有的,其他的都是同样的 5 项: kCollectorTypeMS、kCollectorTypeCMS、kCollectorTypeSS、kCollectorTypeGSS 和 kCollectorTypeCC,所以无论通过编译的配置文件设置,还是通过编译选项进行设置,GC 的选项最终都在上述范围内。

7.3.2 不同类型回收器的实现

介绍完回收器的类型之后,需要关注的是这些回收器的具体实现。在 ART 中,回收器的实现代码位于 art/Runtime/gc/collector 目录之下。回收器本身有一个实现类 GarbageCollector,不同回收方案所对应的回收器是作为 GarbageCollector 的子类进行实现的。

ART GC 的实现代码主要位于 art/Runtime/gc 目录之下。GarbageCollector 类的声明和实现位于 art/Runtime/gc/collector 目录的 garbage_collector.h 和 garbage_collector.cc 中。GarbageCollector 类继承自 RootVisitor、IsMarkedVisitor 和 MarkObjectVisitor，这 3 个类都是在 GC 中会用到的类。GarbageCollector 类中有一个 ScopedPause 类，ScopedPause 类有一个比较重要的成员变量 collector_，它有一个要指向回收器的指针；GarbageCollector 类中比较重要的成员函数是 Run()函数和 RunPhases()函数，Run()函数用于运行回收器，RunPhases()函数用于运行 GC 的所有 phase，代码如下：

```
//第7章/ garbage_collector.h
class GarbageCollector : public RootVisitor, public IsMarkedVisitor,
        public MarkObjectVisitor {
public:
  class SCOPED_LOCKABLE ScopedPause {
   public:
    explicit ScopedPause(GarbageCollector* collector,
        bool with_reporting = true)
        EXCLUSIVE_LOCK_FUNCTION(Locks::mutator_lock_);
    ~ScopedPause() UNLOCK_FUNCTION();

   private:
    const uint64_t start_time_;
    GarbageCollector* const collector_;
    bool with_reporting_;
  };

…
  //Run the garbage collector.
  void Run(GcCause gc_cause, bool clear_soft_references) REQUIRES(!pause_histogram_lock_);
…
protected:
  //Run all of the GC phases.
  virtual void RunPhases() = 0;
…
```

其中，GarbageCollector::Run()函数的实现位于 garbage_collector.cc 中，它调用了 RunPhases()函数。也就是说，GarbageCollector::Run()函数在实现其垃圾回收的功能时通过调用 RunPhases()函数运行所有的 GC phase 来最终实现自己的目标。

GarbageCollector::Run()函数的代码如下：

```
//第7章/ garbage_collector.cc
void GarbageCollector::Run(GcCause gc_cause, bool clear_soft_references)
{
  ScopedTrace trace(android::base::StringPrintf("%s %s GC",
```

```cpp
    PrettyCause(gc_cause), GetName()));
Thread* self = Thread::Current();
uint64_t start_time = NanoTime();
uint64_t thread_cpu_start_time = ThreadCpuNanoTime();
GetHeap()->CalculatePreGcWeightedAllocatedBytes();
Iteration* current_iteration = GetCurrentIteration();
current_iteration->Reset(gc_cause, clear_soft_references);
//Note transaction mode is single-threaded and there's no asynchronous
//GC and this flag doesn't change in the middle of a GC
is_transaction_active_ = Runtime::Current()->IsActiveTransaction();
RunPhases(); //Run all the GC phases.
GetHeap()->CalculatePostGcWeightedAllocatedBytes();
//Add the current timings to the cumulative timings
cumulative_timings_.AddLogger(*GetTimings());
//Update cumulative statistics with how many Bytes the GC iteration
//freed
total_freed_objects_ += current_iteration->GetFreedObjects() +
    current_iteration->GetFreedLargeObjects();
int64_t freed_Bytes = current_iteration->GetFreedBytes() +
    current_iteration->GetFreedLargeObjectBytes();
total_freed_Bytes_ += freed_Bytes;
//Rounding negative freed Bytes to 0 as we are not interested in such
//corner cases
freed_Bytes_histogram_.AddValue(std::max<int64_t>(freed_Bytes / KB,
    0));
uint64_t end_time = NanoTime();
uint64_t thread_cpu_end_time = ThreadCpuNanoTime();
total_thread_cpu_time_ns_ += thread_cpu_end_time -
    thread_cpu_start_time;
current_iteration->SetDurationNs(end_time - start_time);
if (Locks::mutator_lock_->IsExclusiveHeld(self)) {
    //The entire GC was paused, clear the fake pauses which might be in
    //the pause times and add the whole GC duration
    current_iteration->pause_times_.clear();
    RegisterPause(current_iteration->GetDurationNs());
}
total_time_ns_ += current_iteration->GetDurationNs();
for (uint64_t pause_time : current_iteration->GetPauseTimes()) {
    MutexLock mu(self, pause_histogram_lock_);
    pause_histogram_.AdjustAndAddValue(pause_time);
}
is_transaction_active_ = false;
}
```

RunPhases()函数在 GarbageCollector 类中并没有具体实现,要在其子类中根据需求去实现,所以 GarbageCollector 类的子类的 RunPhases()函数是其所对应的回收器的核心实现。GarbageCollector 类的 RunPhases()函数的代码如下:

```
//第7章/ garbage_collector.h
protected:
    //Run all of the GC phases
    virtual void RunPhases() = 0;
```

GarbageCollector类有3个子类：ConcurrentCopying、MarkSweep和SemiSpace。

ConcurrentCopying类的声明和实现位于art/Runtime/gc/collector/目录下的concurrent_copying.h、concurrent_copying-inl.h和concurrent_copying.cc中。它是并发复制回收方案的回收器。

MarkSweep类的声明和实现位于art/Runtime/gc/collector/目录下的mark_sweep.h、mark_sweep-inl.h和mark_sweep.cc中。MarkSweep类还有一个子类，叫作PartialMarkSweep，它的声明和实现位于art/Runtime/gc/collector/目录下的partial_mark_sweep.h和partial_mark_sweep.cc中。PartialMarkSweep类也有个子类，叫作StickyMarkSweep，它的声明和实现位于art/Runtime/gc/collector/目录下的stick_mark_sweep.h和stick_mark_sweep.cc中。

SemiSpace类的声明和实现位于art/Runtime/gc/collector/目录下的semi_space.h、semi_space-inl.h和semi_space.cc中。

上述这些类之间的继承关系如图7.1所示。

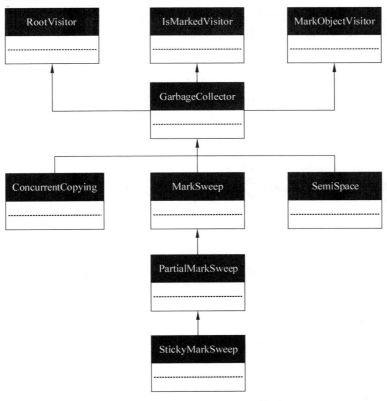

图7.1 回收器类之间的继承关系

图 7.1 展示了回收器的相关类，并且表明了它们之间的继承关系。这些回收器的实现类还需要和前文介绍的回收器的类型对应起来，这样才能构成一个完整的体系。前文所介绍的回收器的类型除了后台专用的 kCollectorTypeHomogeneousSpaceCompact，主要还有 kCollectorTypeMS、kCollectorTypeCMS、kCollectorTypeSS、kCollectorTypeGSS 和 kCollectorTypeCC。

kCollectorTypeMS 类型和 kCollectorTypeCMS 类型的回收器所对应的实现都在 MarkSweep 类、PartialMarkSweep 类和 StickyMarkSweep 类中。MarkSweep 类中的 GetCollectorType() 函数所返回的回收器类型是 kCollectorTypeMS 或者 kCollectorTypeCMS。GetCollectorType() 函数的代码如下：

```
//第 7 章/ mark_sweep.h
CollectorType GetCollectorType() const override {
  return is_concurrent_ ? kCollectorTypeCMS : kCollectorTypeMS;
}
```

GetCollectorType() 函数根据 is_concurrent_ 的值会返回 kCollectorTypeCMS 类型或者 kCollectorTypeMS 类型，前者是并发版本的 MS 回收器类型，后者是普通版本的 MS 回收器类型。PartialMarkSweep 类作为 MarkSweep 的子类，StickyMarkSweep 类作为 PartialMarkSweep 的子类，它们都没有再实现 GetCollectorType() 函数，但都使用了继承自 MarkSweep 类的 GetCollectorType() 函数。由此可以断定，MarkSweep 类、PartialMarkSweep 类和 StickyMarkSweep 类是 kCollectorTypeMS 类型和 kCollectorTypeCMS 类型回收器的实现。

MarkSweep 类、PartialMarkSweep 类和 StickyMarkSweep 类 3 者之间的差别主要是回收策略的差别。回收策略的类型可以从 art/Runtime/gc/collector/gc_type.h 中看到，代码如下：

```
//第 7 章/ gc_type.h
//The type of collection to be performed. The ordering of the enum
//matters, it is used to determine which GCs are run first
enum GcType {
  //Placeholder for when no GC has been performed
  kGcTypeNone,
  //Sticky mark bits GC that attempts to only free objects allocated
  //since the last GC
  kGcTypeSticky,
  //Partial GC that marks the application heap but not the Zygote
  kGcTypePartial,
  //Full GC that marks and frees in both the application and Zygote
  //heap
  kGcTypeFull,
  //Number of different GC types
  kGcTypeMax,
};
```

主要有 3 种回收策略：kGcTypeSticky、kGcTypePartial 和 kGcTypeFull。其中，kGcTypeSticky 表示 GC 只会去尝试释放上次 GC 之后新分配的对象；kGcTypePartial 表示 GC 将会回收应用程序的堆，但是不会回收 Zygote 的堆；kGcTypeFull 表示将会进行完全 GC，回收应用程序和 Zygote 的堆。这 3 种策略对应到具体实现的 MarkSweep 类、PartialMarkSweep 类和 StickyMarkSweep 类就是 kGcTypeSticky 策略对应着 StickyMarkSweep 类，kGcTypePartial 策略对应着 PartialMarkSweep 类，kGcTypeFull 策略对应着 MarkSweep 类。这个对应关系可以从这 3 个类的 GetGcType() 函数的返回值看出。这 3 个类的 GetGcType() 函数的代码如下：

```
//第7章/ mark_sweep.h
//art/Runtime/gc/collector/ mark_sweep.h
  GcType GetGcType() const override {
    return kGcTypeFull;
  }

//第7章/partial_mark_sweep.h
//art/Runtime/gc/collector/partial_mark_sweep.h
  GcType GetGcType() const override {
    return kGcTypePartial;
  }

//第7章/sticky_mark_sweep.h
//art/Runtime/gc/collector/sticky_mark_sweep.h
  GcType GetGcType() const override {
    return kGcTypeSticky;
  }
```

已知 MarkSweep 类、PartialMarkSweep 类和 StickyMarkSweep 类的实现对应着 3 种类型的回收策略：kGcTypeFull、kGcTypePartial 和 kGcTypeSticky，对应着 kCollectorTypeMS 类型和 kCollectorTypeCMS 类型的回收器。在这 3 个类内部，区分是 kCollectorTypeMS 类型还是 kCollectorTypeCMS 类型的实现主要靠 MarkSweep 类中的 RunPhases() 函数，代码如下：

```
//第7章/ mark_sweep.cc
void MarkSweep::RunPhases() {
  Thread* self = Thread::Current();
  InitializePhase();
  Locks::mutator_lock_->AssertNotHeld(self);
  if (IsConcurrent()) {
    GetHeap()->PreGcVerification(this);
    {
      ReaderMutexLock mu(self, *Locks::mutator_lock_);
```

```
      MarkingPhase();
    }
    ScopedPause pause(this);
    GetHeap()->PrePauseRosAllocVerification(this);
    PausePhase();
    RevokeAllThreadLocalBuffers();
  } else {
    ScopedPause pause(this);
    GetHeap()->PreGcVerificationPaused(this);
    MarkingPhase();
    GetHeap()->PrePauseRosAllocVerification(this);
    PausePhase();
    RevokeAllThreadLocalBuffers();
  }
  {
    //Sweeping always done concurrently, even for non concurrent mark
    //sweep
    ReaderMutexLock mu(self, *Locks::mutator_lock_);
    ReclaimPhase();
  }
  GetHeap()->PostGcVerification(this);
  FinishPhase();
}
```

区分是 kCollectorTypeMS 类型还是 kCollectorTypeCMS 类型的实现靠 IsConcurrent()函数的返回值是否为真进行选择。PartialMarkSweep 类和 StickyMarkSweep 类并没有对 RunPhases()函数进行重新实现,所以都采用了 MarkSweep 类中的 RunPhases()函数。至此,可以看出 MarkSweep 类、PartialMarkSweep 类和 StickyMarkSweep 类与两种回收器类型和三种回收策略的对应关系。

kCollectorTypeSS 类型的 kCollectorTypeGSS 类型的回收器对应的实现是 SemiSpace 类,SemiSpace 类对应的回收策略的是 kGcTypePartial 回收策略。这些可以从 SemiSpace 类的 GetCollectorType()函数和 GetGcType()函数看出来,这两个函数的代码如下:

```
//第 7 章/ semi_space.h
GcType GetGcType() const override {
    return kGcTypePartial;
}
CollectorType GetCollectorType() const override {
    return generational_ ? kCollectorTypeGSS : kCollectorTypeSS;
}
```

根据 generational_ 的值可以确定实现的回收器类型是 kCollectorTypeGSS 或 kCollectorTypeSS。这个判断不光用在这里,在 SemiSpace 类的 InitializePhase()、

MarkingPhase()、ReclaimPhase()和 FinishPhase()函数中,都会根据 generational_的值,然后进行不同的处理,以达到区分 kCollectorTypeGSS 或 kCollectorTypeSS 类型处理器的结果。SemiSpace 类的 InitializePhase()、MarkingPhase()、ReclaimPhase()和 FinishPhase()函数正是 SemiSpace 类的核心函数 RunPhases()中的主要组成部分,代码如下:

```
//第7章/ semi_space.cc
void SemiSpace::RunPhases() {
  Thread* self = Thread::Current();
  InitializePhase();
  //Semi-space collector is special since it is sometimes called with
  //the mutators suspended during the zygote creation and collector
  //transitions. If we already exclusively hold the
  //mutator lock, then we can't lock it again since it will cause a
  //deadlock
  if (Locks::mutator_lock_->IsExclusiveHeld(self)) {
    GetHeap()->PreGcVerificationPaused(this);
    GetHeap()->PrePauseRosAllocVerification(this);
    MarkingPhase();
    ReclaimPhase();
    GetHeap()->PostGcVerificationPaused(this);
  } else {
    Locks::mutator_lock_->AssertNotHeld(self);
    {
      ScopedPause pause(this);
      GetHeap()->PreGcVerificationPaused(this);
      GetHeap()->PrePauseRosAllocVerification(this);
      MarkingPhase();
    }
    {
      ReaderMutexLock mu(self, *Locks::mutator_lock_);
      ReclaimPhase();
    }
    GetHeap()->PostGcVerification(this);
  }
  FinishPhase();
}
```

kCollectorTypeCC 类型的回收器对应的实现是 ConcurrentCopying 类,ConcurrentCopying 类对应着 kGcTypeSticky 或 kGcTypePartial 这两种回收策略之一。依然从 ConcurrentCopying 类的 GetCollectorType()函数和 GetGcType()函数获取信息,这两个函数的代码如下:

```
//第7章/concurrent_copying.h
  GcType GetGcType() const override {
```

```
        return (use_generational_cc_ && young_gen_)
            ? kGcTypeSticky
            : kGcTypePartial;
    }
    CollectorType GetCollectorType() const override {
      return kCollectorTypeCC;
    }
```

表达式 use_generational_cc_ && young_gen_ 的值，最终决定了采用哪种回收策略，这个表达式的值也在回收器的具体代码实现中，为不同的回收策略对应了不同的实现。

总结起来，各个回收器的具体实现，重要的是去分析其 RunPhases() 函数的实现，根据其 RunPhases() 函数中所调用的不同内容，实现最终的回收器。同时，在回收器类型、回收策略和回收器的实现类之间，不同的回收策略往往对应着不同的类（ConcurrentCopying 类除外）；相近却不同的回收器类型往往是在类的内部根据某个变量来选择不同的运行路径进行实现，并不会分割出单独的类。回收器类型、回收器实现类和回收策略的对应关系如表 7.1 所示。

表 7.1 回收器类型、回收器实现类和回收策略的对应关系

序号	回收器类型	回收器实现类	回收策略
1	kCollectorTypeMS/ kCollectorTypeCMS	MarkSweep	kGcTypeFull
2		PartialMarkSweep	kGcTypePartial
3		StickyMarkSweep	kGcTypeSticky
4	kCollectorTypeGSS/kCollectorTypeSS	SemiSpace	kGcTypePartial
5	kCollectorTypeCC	ConcurrentCopying	kGcTypeSticky/ kGcTypePartial

本部分对 GC 回收器的类型及不同类型的回收器的实现进行了分析，对于不同回收器的实现，介绍了其具体的实现类及其实现的策略。对于具体的实现类的内部代码及策略的算法并未进行过多探讨，读者可以根据自己的需要选择代码进行深入阅读。

7.4 ART GC 的分配器实现

分配器和回收器是一组相对的概念，GC 回收方案不仅涉及回收器，还涉及分配器。本部分内容将对 ART GC 分配器的实现进行介绍，具体会分为分配器与空间类、分配器与回收器两部分进行介绍。

7.4.1 分配器与空间类

分析 ART GC 的分配器，首先需要了解分配器的类型。ART GC 中所涉及的分配器的类型是通过一个枚举类型 AllocatorType 定义的。它的实现位于 art/Runtime/gc/allocator_ type.h中,代码如下:

```
//第7章/allocator_type.h
//Different types of allocators
//Those marked with * have fast path entrypoints callable from generated
//code
enum AllocatorType {
    //BumpPointer spaces are currently only used for ZygoteSpace
    //construction
    //Use global CAS-based BumpPointer allocator. (*)
    kAllocatorTypeBumpPointer,
    //Use TLAB allocator within BumpPointer space. (*)
    kAllocatorTypeTLAB,
    //Use RosAlloc (segregated size, free list) allocator. (*)
    kAllocatorTypeRosAlloc,
    //Use dlmalloc (well-known C malloc) allocator. (*)
    kAllocatorTypeDlMalloc,
    //Special allocator for non moving objects
    kAllocatorTypeNonMoving,
    //Large object space
    kAllocatorTypeLOS,
    //The following differ from the BumpPointer allocators primarily in
    //that memory is allocated from multiple regions, instead of a single
    //contiguous space.
    //Use CAS-based contiguous bump-pointer allocation within a region
    //(*)
    kAllocatorTypeRegion,
    //Use region pieces as TLABs. Default for most small objects. (*)
    kAllocatorTypeRegionTLAB,
};
```

这些分配器的类型的具体介绍,在其对应的注释中有对应的详细介绍。分配器类型的初始值在 Heap 类的构造函数中进行了初始化,可以在 art/Runtime/gc/heap.cc 中看到相关代码,具体如下:

```
//第7章/heap.cc
    current_allocator_(kAllocatorTypeDlMalloc),
    current_non_moving_allocator_(kAllocatorTypeNonMoving),
```

其中的 current_allocator_ 是 AllocatorType 类型的,被设置为 kAllocatorTypeDlMalloc;current_non_moving_allocator_ 是 const AllocatorType 类型的,被设置为 kAllocatorTypeNonMoving。current_allocator_ 和 current_non_moving_allocator_ 的声明位于 art/Runtime/gc/heap.h 中,代码如下:

```
//第7章/heap.h
    //Allocator type
    AllocatorType current_allocator_;
    const AllocatorType current_non_moving_allocator_;
```

不同的分配器,对应着不同的内存空间,而不同的内存空间,又都有着自身的内存分配方式。ART GC 中所涉及的内存空间是用一个枚举类型 SpaceType 来表示的,它位于 art/Runtime/gc/space/space.h 文件中,代码如下:

```
//第7章/space.h
enum SpaceType {
    kSpaceTypeImageSpace,
    kSpaceTypeMallocSpace,
    kSpaceTypeZygoteSpace,
    kSpaceTypeBumpPointerSpace,
    kSpaceTypeLargeObjectSpace,
    kSpaceTypeRegionSpace,
};
```

实际上,这些内存空间的类型都有对应的类进行实现,实现类之间也有一定的关系。每种空间内部都有自己的内存分配方式,每种空间的实现类也都有自己的类型。这里的内存分配空间的实现类的代码都位于 art/Runtime/gc/space 目录中。

从代码实现上讲,可以看到的内存分配空间类有 Space、ImageSpace、MallocSpace、DlMallocSpace、RosAllocSpace、ZygoteSpace、BumpPointerSpace、LargeObjectSpace 和 RegionSpace 等,这些类都有对应的.h 和.cc 文件用于专门实现,部分类还有对应的-inl.h 文件用于实现其内联函数。在这几个类中,Space 是这一系列空间类的基类;ImageSpace 类是 kSpaceTypeImageSpace 类型的空间;MallocSpace 类是 kSpaceTypeMallocSpace 类型的空间,同时 DlMallocSpace 和 RosAllocSpace 都是它的子类;ZygoteSpace 类是 kSpaceTypeZygoteSpace 类型的空间;BumpPointerSpace 类是 kSpaceTypeBumpPointerSpace 类型的空间;LargeObjectSpace 类是 kSpaceTypeLargeObjectSpace 类型的空间;RegionSpace 类是 kSpaceTypeRegionSpace 类型的空间。每个类都可以从其 GetType 成员函数获取其类型,个别类如 DlMallocSpace 和 RosAllocSpace 没有实现该函数,但它们继承了父类的该函数。

空间类有继承关系,其具体的关系如图 7.2 所示。

既然这些空间类本身具备了内存分配能力,那么如何和多种类型的分配器对应起来呢?这就是下一步需要考虑的问题。在 art/Runtime/gc/heap-inl.h 文件的 Heap::TryToAllocate()函数中,可以看到在实际分配空间的时候,不同类型的分配器是如何与空间类及其分配空间配合使用的。Heap::TryToAllocate()函数的代码如下:

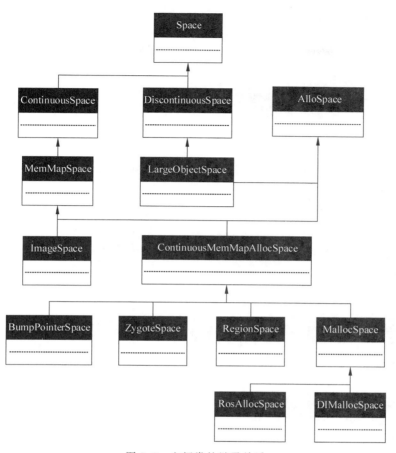

图 7.2 空间类的继承关系

```
//第 7 章/heap - inl.h
template < const bool kInstrumented, const bool kGrow >
inline mirror::Object * Heap::TryToAllocate(Thread * self,
    AllocatorType allocator_type, size_t alloc_size,
    size_t * Bytes_allocated, size_t * usable_size,
    size_t * Bytes_tl_bulk_allocated) {
  if (allocator_type != kAllocatorTypeRegionTLAB &&
      allocator_type != kAllocatorTypeTLAB &&
      allocator_type != kAllocatorTypeRosAlloc &&
      UNLIKELY(IsOutOfMemoryOnAllocation(allocator_type,
                                  alloc_size, kGrow))) {
    return nullptr;
  }
  mirror::Object * ret;
  switch (allocator_type) {
```

```cpp
case kAllocatorTypeBumpPointer: {
  DCHECK(bump_pointer_space_ != nullptr);
  alloc_size = RoundUp(alloc_size,
                       space::BumpPointerSpace::kAlignment);
  ret = bump_pointer_space_->AllocNonvirtual(alloc_size);
  if (LIKELY(ret != nullptr)) {
    *Bytes_allocated = alloc_size;
    *usable_size = alloc_size;
    *Bytes_tl_bulk_allocated = alloc_size;
  }
  break;
}
case kAllocatorTypeRosAlloc: {
  if (kInstrumented && UNLIKELY(is_running_on_memory_tool_)) {
    //If running on ASan, we should be using the instrumented path
    size_t max_Bytes_tl_bulk_allocated =
        rosalloc_space_->MaxBytesBulkAllocatedFor(alloc_size);
    if (UNLIKELY(IsOutOfMemoryOnAllocation(allocator_type,
        max_Bytes_tl_bulk_allocated, kGrow))) {
      return nullptr;
    }
    ret = rosalloc_space_->Alloc(self, alloc_size, Bytes_allocated,
                                 usable_size,
                                 Bytes_tl_bulk_allocated);
  } else {
    DCHECK(!is_running_on_memory_tool_);
    size_t max_Bytes_tl_bulk_allocated =
        rosalloc_space_->MaxBytesBulkAllocatedForNonvirtual(
            alloc_size);
    if (UNLIKELY(IsOutOfMemoryOnAllocation(allocator_type,
        max_Bytes_tl_bulk_allocated, kGrow))) {
      return nullptr;
    }
    if (!kInstrumented) {
      DCHECK(!rosalloc_space_->CanAllocThreadLocal(self, alloc_size));
    }
    ret = rosalloc_space_->AllocNonvirtual(self,
                                           alloc_size,
                                           Bytes_allocated,
                                           usable_size,
                                           Bytes_tl_bulk_allocated);
  }
  break;
}
case kAllocatorTypeDlMalloc: {
  if (kInstrumented && UNLIKELY(is_running_on_memory_tool_)) {
```

```cpp
            //If running on ASan, we should be using the instrumented path
            ret = dlmalloc_space_->Alloc(self,
                                         alloc_size,
                                         Bytes_allocated,
                                         usable_size,
                                         Bytes_tl_bulk_allocated);
        } else {
            DCHECK(!is_running_on_memory_tool_);
            ret = dlmalloc_space_->AllocNonvirtual(self,
                                                   alloc_size,
                                                   Bytes_allocated,
                                                   usable_size,
                                                   Bytes_tl_bulk_allocated);
        }
        break;
    }
    case kAllocatorTypeNonMoving: {
        ret = non_moving_space_->Alloc(self,
                                       alloc_size,
                                       Bytes_allocated,
                                       usable_size,
                                       Bytes_tl_bulk_allocated);
        break;
    }
    case kAllocatorTypeLOS: {
        ret = large_object_space_->Alloc(self,
                                         alloc_size,
                                         Bytes_allocated,
                                         usable_size,
                                         Bytes_tl_bulk_allocated);
        //Note that the bump pointer spaces aren't necessarily next to
        //the other continuous spaces like the non-moving alloc space or
        //the zygote space
        DCHECK(ret == nullptr || large_object_space_->Contains(ret));
        break;
    }
    case kAllocatorTypeRegion: {
        DCHECK(region_space_ != nullptr);
        alloc_size = RoundUp(alloc_size, space::RegionSpace::kAlignment);
        ret = region_space_->AllocNonvirtual<false>(alloc_size,
            Bytes_allocated, usable_size, Bytes_tl_bulk_allocated);
        break;
    }
    case kAllocatorTypeTLAB:
        FALLTHROUGH_INTENDED;
    case kAllocatorTypeRegionTLAB: {
```

```
            DCHECK_ALIGNED(alloc_size, kObjectAlignment);
            static_assert(space::RegionSpace::kAlignment ==
                          space::BumpPointerSpace::kAlignment,
                          "mismatched alignments");
            static_assert(kObjectAlignment ==
                          space::BumpPointerSpace::kAlignment,
                          "mismatched alignments");
            if (UNLIKELY(self->TlabSize() < alloc_size)) {
              //kAllocatorTypeTLAB may be the allocator for region space TLAB
              //if the GC is not marking, that is why the allocator is not
              //passed down
              return AllocWithNewTLAB(self,
                                     alloc_size,
                                     kGrow,
                                     Bytes_allocated,
                                     usable_size,
                                     Bytes_tl_bulk_allocated);
            }
            //The allocation can't fail
            ret = self->AllocTlab(alloc_size);
            DCHECK(ret != nullptr);
            *Bytes_allocated = alloc_size;
            *Bytes_tl_bulk_allocated = 0; //Allocated in an existing buffer
            *usable_size = alloc_size;
            break;
          }
          default: {
            LOG(FATAL) << "Invalid allocator type";
            ret = nullptr;
          }
        }
        return ret;
      }
```

这里用了一个 switch-case 结构，根据分配器类型 allocator_type 的不同取值，去选择对应的空间类，并调用其分配空间函数，以此去实现内存分配。kAllocatorTypeBumpPointer 类型的分配器调用的是 bump_pointer_space_->AllocNonvirtual()函数；kAllocatorTypeRosAlloc 类型的分配器调用的是 rosalloc_space_->Alloc()函数或 rosalloc_space_->AllocNonvirtual()函数；kAllocatorTypeDlMalloc 类型的分配器调用的是 dlmalloc_space_->Alloc()函数或 dlmalloc_space_->AllocNonvirtual()函数；kAllocatorTypeNonMoving 类型的分配器调用的是 non_moving_space_->Alloc()函数；kAllocatorTypeLOS 类型的分配器调用的是 large_object_space_->Alloc()函数；kAllocatorTypeRegion 类型的分配器调用的是 region_space_->AllocNonvirtual()函数；

kAllocatorTypeTLAB 类型和 kAllocatorTypeRegionTLAB 类型的分配器调用的是 AllocWithNewTLAB()函数或 self->AllocTlab()函数。

因为 bump_pointer_space_、rosalloc_space_ 和 dlmalloc_space_ 等都是 Heap 类的成员变量，所以在 art/Runtime/gc/heap.h 中都可以找到其类型，它们的类型都是某个空间类的指针。将分配器类型、对应的空间类、类中分配空间的函数列成一个表，可以很清楚地看到分配器与空间类及分配空间函数的对应关系，具体如表 7.2 所示。

表 7.2 分配器与空间类及分配空间函数的对应关系

序号	分配器类型	空间类	分配空间函数
1	kAllocatorTypeBumpPointer	BumpPointerSpace	AllocNonvirtual()函数
2	kAllocatorTypeRosAlloc	RosAllocSpace	Alloc()函数或 AllocNonvirtual()函数
3	kAllocatorTypeDlMalloc	DlMallocSpace	Alloc()函数或 AllocNonvirtual()函数
4	kAllocatorTypeNonMoving	MallocSpace	Alloc()函数
5	kAllocatorTypeLOS	LargeObjectSpace	Alloc()函数
6	kAllocatorTypeRegion	RegionSpace	AllocNonvirtual()函数
7	kAllocatorTypeTLAB		AllocWithNewTLAB()函数或 Thread:: AllocTlab()函数
8	kAllocatorTypeRegionTLAB		

如表 7.2 所示，所有的分配器类型都有对应的分配空间的函数为其分配空间。这就理清了分配器类型与空间类及其函数的关系，也理清了分配器的主要实现。

表 7.2 中的第 7 项和第 8 项可以进一步深入跟踪，在 AllocWithNewTLAB()函数内部，kAllocatorTypeTLAB 类型的分配器会调用 BumpPointerSpace 类的 AllocNewTlab()函数；kAllocatorTypeRegionTLAB 类型的分配器会调用 RegionSpace 类的 AllocNewTlab()函数或者 AllocNonvirtual()函数，但在某些条件得不到满足的情况下也会调用 Thread:: AllocTlab()函数。AllocWithNewTLAB()函数位于 art/Runtime/gc/heap.cc 中，代码如下：

```
//第7章/heap.cc
mirror::Object * Heap::AllocWithNewTLAB(Thread * self,
                                        size_t alloc_size,
                                        bool grow,
                                        size_t * Bytes_allocated,
                                        size_t * usable_size,
                                        size_t * Bytes_tl_bulk_allocated) {
    const AllocatorType allocator_type = GetCurrentAllocator();
    if (kUsePartialTlabs && alloc_size <= self->TlabRemainingCapacity()) {
        DCHECK_GT(alloc_size, self->TlabSize());
        //There is enough space if we grow the TLAB. Lets do that. This
        //increases the TLAB Bytes
        const size_t min_expand_size = alloc_size - self->TlabSize();
        const size_t expand_Bytes = std::max(
            min_expand_size,
```

```cpp
                    std::min(self->TlabRemainingCapacity() - self->TlabSize(), kPartialTlabSize));
  if (UNLIKELY(IsOutOfMemoryOnAllocation(allocator_type,
                                         expand_Bytes, grow))) {
    return nullptr;
  }
  *Bytes_tl_bulk_allocated = expand_Bytes;
  self->ExpandTlab(expand_Bytes);
  DCHECK_LE(alloc_size, self->TlabSize());
} else if (allocator_type == kAllocatorTypeTLAB) {
  DCHECK(bump_pointer_space_ != nullptr);
  const size_t new_tlab_size = alloc_size + kDefaultTLABSize;
  if (UNLIKELY(IsOutOfMemoryOnAllocation(allocator_type,
                                         new_tlab_size, grow))) {
    return nullptr;
  }
  //Try allocating a new thread local buffer, if the allocation fails
  //the space must be full so return null
  if (!bump_pointer_space_->AllocNewTlab(self, new_tlab_size)) {
    return nullptr;
  }
  *Bytes_tl_bulk_allocated = new_tlab_size;
} else {
  DCHECK(allocator_type == kAllocatorTypeRegionTLAB);
  DCHECK(region_space_ != nullptr);
  if (space::RegionSpace::kRegionSize >= alloc_size) {
    //Non-large. Check OOME for a tlab
    if (LIKELY(!IsOutOfMemoryOnAllocation(allocator_type,
         space::RegionSpace::kRegionSize, grow))) {
      const size_t new_tlab_size = kUsePartialTlabs
          ? std::max(alloc_size, kPartialTlabSize)
          : gc::space::RegionSpace::kRegionSize;
      //Try to allocate a tlab
      if (!region_space_->AllocNewTlab(self, new_tlab_size)) {
        //Failed to allocate a tlab. Try non-tlab
        return region_space_->AllocNonvirtual<false>(alloc_size,
            Bytes_allocated, usable_size, Bytes_tl_bulk_allocated);
      }
      *Bytes_tl_bulk_allocated = new_tlab_size;
      //Fall-through to using the TLAB below
    } else {
      //Check OOME for a non-tlab allocation
      if (!IsOutOfMemoryOnAllocation(allocator_type,
                                     alloc_size, grow)) {
        return region_space_->AllocNonvirtual<false>(alloc_size,
            Bytes_allocated, usable_size, Bytes_tl_bulk_allocated);
      }
```

```
            //Neither tlab or non-tlab works. Give up
            return nullptr;
        }
    } else {
        //Large. Check OOME
        if (LIKELY(!IsOutOfMemoryOnAllocation(allocator_type, alloc_size,
                                             grow))) {
            return region_space_ -> AllocNonvirtual<false>(alloc_size,
                Bytes_allocated, usable_size, Bytes_tl_bulk_allocated);
        }
        return nullptr;
    }
}
//Refilled TLAB, return
mirror::Object* ret = self -> AllocTlab(alloc_size);
DCHECK(ret != nullptr);
*Bytes_allocated = alloc_size;
*usable_size = alloc_size;
return ret;
}
```

根据上述分析,可以对表 7.2 进行更新,具体如表 7.3 所示。

表 7.3 分配器与空间类及分配空间函数的对应关系(更新版)

序号	分配器类型	空间类	分配空间函数
1	kAllocatorTypeBumpPointer	BumpPointerSpace	AllocNonvirtual()函数
2	kAllocatorTypeRosAlloc	RosAllocSpace	Alloc()函数或 AllocNonvirtual()函数
3	kAllocatorTypeDlMalloc	DlMallocSpace	Alloc()函数或 AllocNonvirtual()函数
4	kAllocatorTypeNonMoving	MallocSpace	Alloc()函数
5	kAllocatorTypeLOS	LargeObjectSpace	Alloc()函数
6	kAllocatorTypeRegion	RegionSpace	AllocNonvirtual()函数
7	kAllocatorTypeTLAB	BumpPointerSpace	AllocNewTlab()函数或 Thread::AllocTlab()函数
8	kAllocatorTypeRegionTLAB	RegionSpace	AllocNewTlab()函数或 AllocNonvirtual()函数或 Thread::AllocTlab()函数

分配器涉及的代码目录,除了 art/Runtime/gc/space 目录,还有 art/Runtime/gc/allocator 目录。在 art/Runtime/gc/allocator/目录中,还有 dlmalloc.h、dlmalloc.cc、rosalloc.h、rosalloc-inl.h 和 rosalloc.cc 这 5 个文件。dlmalloc.h 和 dlmalloc.cc 是 kAllocatorTypeDlMalloc 类型分配器分配空间时的回调函数。rosalloc.h、rosalloc-inl.h 和 rosalloc.cc 是 kAllocatorTypeRosAlloc 类型分配器中真正分配空间的算法实现,是通过 RosAllocSpace 的相关函数去调用的。

7.4.2 分配器与回收器

分配器除了有自己的实现和空间类的问题,还涉及回收器的匹配问题,这个匹配在 art/Runtime/gc/heap.cc 中的 Heap::ChangeCollector()函数的实现中有体现,具体的代码如下:

```
//第7章/heap.cc
void Heap::ChangeCollector(CollectorType collector_type) {
  //TODO: Only do this with all mutators suspended to avoid races
  if (collector_type != collector_type_) {
    collector_type_ = collector_type;
    gc_plan_.clear();
    switch (collector_type_) {
      case kCollectorTypeCC: {
        if (use_generational_cc_) {
          gc_plan_.push_back(collector::kGcTypeSticky);
        }
        gc_plan_.push_back(collector::kGcTypeFull);
        if (use_tlab_) {
          ChangeAllocator(kAllocatorTypeRegionTLAB);
        } else {
          ChangeAllocator(kAllocatorTypeRegion);
        }
        break;
      }
      case kCollectorTypeSS: //Fall-through
      case kCollectorTypeGSS: {
        gc_plan_.push_back(collector::kGcTypeFull);
        if (use_tlab_) {
          ChangeAllocator(kAllocatorTypeTLAB);
        } else {
          ChangeAllocator(kAllocatorTypeBumpPointer);
        }
        break;
      }
      case kCollectorTypeMS: {
        gc_plan_.push_back(collector::kGcTypeSticky);
        gc_plan_.push_back(collector::kGcTypePartial);
        gc_plan_.push_back(collector::kGcTypeFull);
        ChangeAllocator(kUseRosAlloc ? kAllocatorTypeRosAlloc :
                                       kAllocatorTypeDlMalloc);
        break;
      }
      case kCollectorTypeCMS: {
        gc_plan_.push_back(collector::kGcTypeSticky);
        gc_plan_.push_back(collector::kGcTypePartial);
```

```
        gc_plan_.push_back(collector::kGcTypeFull);
        ChangeAllocator(kUseRosAlloc ? kAllocatorTypeRosAlloc :
                                       kAllocatorTypeDlMalloc);
        break;
      }
      default: {
        UNIMPLEMENTED(FATAL);
        UNREACHABLE();
      }
    }
    if (IsGcConcurrent()) {
      concurrent_start_Bytes_ =
          UnsignedDifference(target_footprint_.load(
          std::memory_order_relaxed), kMinConcurrentRemainingBytes);
    } else {
      concurrent_start_Bytes_ = std::numeric_limits<size_t>::max();
    }
  }
}
```

这里采用了一个 switch-case 结构,根据回收器的类型匹配对应的分配器。将分配器与回收器的对应关系用表格的形式表现出来,如表 7.4 所示。

表 7.4 分配器与回收器的对应关系

序号	分配器类型	回收器类型
1	kAllocatorTypeRegionTLAB/ kAllocatorTypeRegion	kCollectorTypeCC
2	kAllocatorTypeTLAB/ kAllocatorTypeBumpPointer	kCollectorTypeSS、kCollectorTypeGSS
3	kAllocatorTypeRosAlloc/ kAllocatorTypeDlMalloc	kCollectorTypeMS
4	kAllocatorTypeRosAlloc/ kAllocatorTypeDlMalloc	kCollectorTypeCMS

至此,分配器与回收器的对应关系也已经显现出来,从表 7.4 中可以看到其对应关系。分配器类型与空间类的对应,以及分配器类型与回收器类型的对应,一起展现了一个 GC 内部的基本框架。本部分内容从分配器的实现及其和空间类、回收器的对应关系入手,为读者展示了一个以分配器为中心的 ART GC 图景。

7.5 ART GC 的使用流程

前文介绍了 ART GC 回收方案及分配器和回收器的实现,本部分内容将从 Heap 类的实现谈起,从中发现分配器和回收器的调用接口,理清分配器和回收器的使用。

在 Heap 类中，有负责保存与分配器和回收器相关的成员变量，也有对其进行操作的相关成员函数。与分配器相关的成员变量在上文已经介绍了。与回收器相关的成员变量主要有 collector_type_、foreground_collector_type_、background_collector_type_、desired_collector_type_ 和 garbage_collectors_，它们分别对应着目前的回收器类型、前台回收器类型、后台回收器类型、期望的回收器类型和回收器列表，代码如下：

```cpp
//第 7 章/heap.h
…
  //The current collector type
  CollectorType collector_type_;
  //Which collector we use when the app is in the foreground
  CollectorType foreground_collector_type_;
  //Which collector we will use when the app is notified of a transition
  //to background
  CollectorType background_collector_type_;
  //Desired collector type, heap trimming daemon transitions the heap if
  //it is != collector_type_
  CollectorType desired_collector_type_;
…
  std::vector<collector::GarbageCollector*> garbage_collectors_;
```

与这些成员变量配套的还有一些成员函数，这些成员函数主要负责获取或者对这些成员变量进行修改。下面将分成分配器的使用和回收器的使用两个部分进行具体介绍。

7.5.1　分配器的使用

分配器要为新建立的对象分配空间，这个操作主要靠 Heap 类的 AllocObjectWithAllocator() 函数实现，AllocObjectWithAllocator() 函数的实现位于 art/Runtime/gc/heap-inl.h 中，代码如下：

```cpp
//第 7 章/heap-inl.h
template <bool kInstrumented, bool kCheckLargeObject,
    typename PreFenceVisitor>
inline mirror::Object* Heap::AllocObjectWithAllocator(Thread* self,
      ObjPtr<mirror::Class> klass, size_t Byte_count,
      AllocatorType allocator, const PreFenceVisitor& pre_fence_visitor) {
  if (kIsDebugBuild) {
    CheckPreconditionsForAllocObject(klass, Byte_count);
    //Since allocation can cause a GC which will need to SuspendAll
    //make sure all allocations are done in the runnable state where
    //suspension is expected
    CHECK_EQ(self->GetState(), kRunnable);
```

```cpp
  self->AssertThreadSuspensionIsAllowable();
  self->AssertNoPendingException();
  //Make sure to preserve klass
  StackHandleScope<1> hs(self);
  HandleWrapperObjPtr<mirror::Class> h = hs.NewHandleWrapper(&klass);
  self->PoisonObjectPointers();
}
//Need to check that we aren't the large object allocator since the
//large object allocation code path includes this function. If we
//didn't check we would have an infinite loop
ObjPtr<mirror::Object> obj;
if (kCheckLargeObject && UNLIKELY(ShouldAllocLargeObject(klass, Byte_count))) {
  obj = AllocLargeObject<kInstrumented, PreFenceVisitor>(self, &klass,
      Byte_count, pre_fence_visitor);
  if (obj != nullptr) {
    return obj.Ptr();
  } else {
    //There should be an OOM exception, since we are retrying, clear
    //it
    self->ClearException();
  }
  //If the large object allocation failed, try to use the normal
  //spaces (main space, non moving space). This can happen if there is
  //significant virtual address space fragmentation
}
//Bytes allocated for the (individual) object
size_t Bytes_allocated;
size_t usable_size;
size_t new_num_Bytes_allocated = 0;
if (IsTLABAllocator(allocator)) {
  Byte_count = RoundUp(Byte_count, space::BumpPointerSpace::kAlignment);
}
//If we have a thread local allocation we don't need to update Bytes
//allocated
if (IsTLABAllocator(allocator) && Byte_count <= self->TlabSize()) {
  obj = self->AllocTlab(Byte_count);
  DCHECK(obj != nullptr) << "AllocTlab can't fail";
  obj->SetClass(klass);
  if (kUseBakerReadBarrier) {
    obj->AssertReadBarrierState();
  }
  Bytes_allocated = Byte_count;
  usable_size = Bytes_allocated;
  pre_fence_visitor(obj, usable_size);
  QuasiAtomic::ThreadFenceForConstructor();
} else if (
```

```cpp
      !kInstrumented && allocator == kAllocatorTypeRosAlloc &&
      (obj = rosalloc_space_->AllocThreadLocal(self, Byte_count,
      &Bytes_allocated)) != nullptr && LIKELY(obj != nullptr)) {
    DCHECK(!is_running_on_memory_tool_);
    obj->SetClass(klass);
    if (kUseBakerReadBarrier) {
      obj->AssertReadBarrierState();
    }
    usable_size = Bytes_allocated;
    pre_fence_visitor(obj, usable_size);
    QuasiAtomic::ThreadFenceForConstructor();
  } else {
    //Bytes allocated that includes bulk thread-local buffer allocations
    //in addition to direct non-TLAB object allocations
    size_t Bytes_tl_bulk_allocated = 0u;
    obj = TryToAllocate<kInstrumented, false>(self, allocator,
        Byte_count, &Bytes_allocated, &usable_size,
        &Bytes_tl_bulk_allocated);
    if (UNLIKELY(obj == nullptr)) {
      //AllocateInternalWithGc can cause thread suspension, if someone
      //instruments the entrypoints or changes the allocator in a suspend
      //point here, we need to retry the allocation
      obj = AllocateInternalWithGc(self,
                                   allocator,
                                   kInstrumented,
                                   Byte_count,
                                   &Bytes_allocated,
                                   &usable_size,
                                   &Bytes_tl_bulk_allocated, &klass);
      if (obj == nullptr) {
        //The only way that we can get a null return if there is no
        //pending exception is if the allocator or instrumentation
        //changed
        if (!self->IsExceptionPending()) {
          //AllocObject will pick up the new allocator type, and
          //instrumented as true is the safe default
          return AllocObject</*kInstrumented=*/true>(self,
                                                      klass,
                                                      Byte_count,
                                                      pre_fence_visitor);
        }
        return nullptr;
      }
    }
    DCHECK_GT(Bytes_allocated, 0u);
    DCHECK_GT(usable_size, 0u);
```

```cpp
    obj->SetClass(klass);
    if (kUseBakerReadBarrier) {
      obj->AssertReadBarrierState();
    }
    if (collector::SemiSpace::kUseRememberedSet && UNLIKELY(
        allocator == kAllocatorTypeNonMoving)) {
      //(Note this if statement will be constant folded away for the
      //fast-path quick entry points.) Because SetClass() has no write
      //barrier, if a non-moving space allocation, we need a write
      //barrier as the class pointer may point to the bump pointer
      //space (where the class pointer is an "old-to-young" reference
      //though rare) under the GSS collector with the remembered set
      //enabled. We don't need this for kAllocatorTypeRosAlloc/DlMalloc
      //cases because we don't directly allocate into the main alloc
      //space (besides promotions) under the SS/GSS collector
      WriteBarrier::ForFieldWrite(obj, mirror::Object::ClassOffset(),
                                  klass);
    }
    pre_fence_visitor(obj, usable_size);
    QuasiAtomic::ThreadFenceForConstructor();
    if (Bytes_tl_bulk_allocated > 0) {
      size_t num_Bytes_allocated_before =
          num_Bytes_allocated_.fetch_add(Bytes_tl_bulk_allocated,
                                         std::memory_order_relaxed);
      new_num_Bytes_allocated = num_Bytes_allocated_before +
                                Bytes_tl_bulk_allocated;
      //Only trace when we get an increase in the number of Bytes
      //allocated. This happens when obtaining a new TLAB and isn't often
      //enough to hurt performance according to golem
      TraceHeapSize(new_num_Bytes_allocated);
    }
  }
  if (kIsDebugBuild && Runtime::Current()->IsStarted()) {
    CHECK_LE(obj->SizeOf(), usable_size);
  }
  //TODO: Deprecate
  if (kInstrumented) {
    if (Runtime::Current()->HasStatsEnabled()) {
      RuntimeStats* thread_stats = self->GetStats();
      ++thread_stats->allocated_objects;
      thread_stats->allocated_Bytes += Bytes_allocated;
      RuntimeStats* global_stats = Runtime::Current()->GetStats();
      ++global_stats->allocated_objects;
      global_stats->allocated_Bytes += Bytes_allocated;
    }
  } else {
```

```cpp
      DCHECK(!Runtime::Current()->HasStatsEnabled());
    }
    if (kInstrumented) {
      if (IsAllocTrackingEnabled()) {
        //allocation_records_ is not null since it never becomes null after
        //allocation tracking is enabled
        DCHECK(allocation_records_ != nullptr);
        allocation_records_->RecordAllocation(self, &obj, Bytes_allocated);
      }
      AllocationListener* l =
          alloc_listener_.load(std::memory_order_seq_cst);
      if (l != nullptr) {
        //Same as above. We assume that a listener that was once stored
        //will never be deleted. Otherwise we'd have to perform this under
        //a lock
        l->ObjectAllocated(self, &obj, Bytes_allocated);
      }
    } else {
      DCHECK(!IsAllocTrackingEnabled());
    }
    if (AllocatorHasAllocationStack(allocator)) {
      PushOnAllocationStack(self, &obj);
    }
    if (kInstrumented) {
      if (gc_stress_mode_) {
        CheckGcStressMode(self, &obj);
      }
    } else {
      DCHECK(!gc_stress_mode_);
    }
    //IsGcConcurrent() isn't known at compile time so we can optimize by
    //not checking it for the BumpPointer or TLAB allocators. This is nice
    //since it allows the entire if statement to be optimized out. And for
    //the other allocators, AllocatorMayHaveConcurrentGC is a constant
    //since the allocator_type should be constant propagated
    if (AllocatorMayHaveConcurrentGC(allocator) && IsGcConcurrent()) {
      //New_num_Bytes_allocated is zero if we didn't update
      //num_Bytes_allocated_. That's fine
      CheckConcurrentGCForJava(self, new_num_Bytes_allocated, &obj);
    }
    VerifyObject(obj);
    self->VerifyStack();
    return obj.Ptr();
}
```

AllocObjectWithAllocator()函数内部也有不同的分支,用于执行不同的分配空间动作,可以跟踪代码继续深入。AllocObjectWithAllocator()函数在使用上通常会被 Class::Alloc()函数(art/Runtime/mirror/class-alloc-inl.h)、Array::Alloc()函数(art/Runtime/mirror/array-alloc-inl.h)和 String::Alloc()函数(art/Runtime/mirror/string-alloc-inl.h)这一类型的函数所调用,都属于为某种类型进行分配空间的函数。

7.5.2 回收器的使用

GC 回收器主要在进行垃圾回收的时候使用,而垃圾回收的执行需要触发事件进行触发。垃圾回收的触发,在实现上专门由一个枚举类型 GcCause 表示,这个枚举类型的定义位于 art/Runtime/gc/gc_cause.h 中,代码如下:

```
//第 7 章/gc_cause.h
//What caused the GC
enum GcCause {
  //Invalid GC cause used as a placeholder
  kGcCauseNone,
  //GC triggered by a failed allocation. Thread doing allocation is
  //blocked waiting for GC before retrying allocation
  kGcCauseForAlloc,
  //A background GC trying to ensure there is free memory ahead of
  //allocations
  kGcCauseBackground,
  //An explicit System.gc() call
  kGcCauseExplicit,
  //GC triggered for a native allocation when
  //NativeAllocationGcWatermark is exceeded. (This may be a blocking GC
  //depending on whether we run a non-concurrent collector)
  kGcCauseForNativeAlloc,
  //GC triggered for a collector transition
  kGcCauseCollectorTransition,
  //Not a real GC cause, used when we disable moving GC (currently for
  //GetPrimitiveArrayCritical)
  kGcCauseDisableMovingGc,
  //Not a real GC cause, used when we trim the heap
  kGcCauseTrim,
  //Not a real GC cause, used to implement exclusion between GC and
  //instrumentation
  kGcCauseInstrumentation,
  //Not a real GC cause, used to add or remove app image spaces
  //kGcCauseAddRemoveAppImageSpace, Not a real GC cause, used to
  //implement exclusion between GC and debugger
  kGcCauseDebugger,
  //GC triggered for background transition when both foreground and
```

```
    //background collector are CMS
    kGcCauseHomogeneousSpaceCompact,
    //Class linker cause, used to guard filling art methods with special
    //values
    kGcCauseClassLinker,
    //Not a real GC cause, used to implement exclusion between code cache
    //metadata and GC
    kGcCauseJitCodeCache,
    //Not a real GC cause, used to add or remove system-weak holders
    kGcCauseAddRemoveSystemWeakHolder,
    //Not a real GC cause, used to prevent hprof running in the middle of
    //GC
    kGcCauseHprof,
    //Not a real GC cause, used to prevent GetObjectsAllocated running in
    //the middle of GC
    kGcCauseGetObjectsAllocated,
    //GC cause for the profile saver
    kGcCauseProfileSaver,
};
```

在上述多种垃圾回收触发中，有很多种并不是真的垃圾回收触发，而是用作特殊用途的，这些触发类型在注释上都表明了 Not a real GC cause。经常使用的主要有 kGcCauseForAlloc、kGcCauseBackground、kGcCauseExplicit、kGcCauseForNativeAlloc 和 kGcCauseCollectorTransition 等。接下来将对 kGcCauseForAlloc、kGcCauseBackground 和 kGcCauseExplicit 这 3 种触发的相关内容使用进行介绍。

1. kGcCauseForAlloc

kGcCauseForAlloc 指的是在分配空间时触发 GC，这种情况下通常是在内存分配失败时触发 GC，在 GC 完成之后再次分配空间。这个过程在代码实现上，首先开始于上文提到的进行内存分配的 AllocObjectWithAllocator() 函数（位于 art/Runtime/gc/heap-inl.h 中），在函数中调用 TryToAllocate() 函数进行内存分配，分配失败的时候会去调用 AllocateInternalWithGc() 函数。AllocateInternalWithGc() 函数的实现位于 art/Runtime/gc/heap.cc 中，它主要通过 CollectGarbageInternal() 函数进行垃圾回收，然后进行内存分配，代码如下：

```
//第 7 章/heap.cc
mirror::Object* Heap::AllocateInternalWithGc(Thread* self,
    AllocatorType allocator, bool instrumented, size_t alloc_size,
    size_t* Bytes_allocated, size_t* usable_size,
    size_t* Bytes_tl_bulk_allocated, ObjPtr<mirror::Class>* klass) {
  bool was_default_allocator = allocator == GetCurrentAllocator();
  //Make sure there is no pending exception since we may need to throw
  //an OOME
```

```cpp
self->AssertNoPendingException();
DCHECK(klass != nullptr);
StackHandleScope<1> hs(self);
HandleWrapperObjPtr<mirror::Class> h(hs.NewHandleWrapper(klass));
//The allocation failed. If the GC is running, block until it
//completes, and then retry the allocation
collector::GcType last_gc = WaitForGcToComplete(kGcCauseForAlloc, self);
//If we were the default allocator but the allocator changed while we
//were suspended, abort the allocation
if ((was_default_allocator && allocator != GetCurrentAllocator()) ||
    (!instrumented && EntrypointsInstrumented())) {
  return nullptr;
}
if (last_gc != collector::kGcTypeNone) {
  //A GC was in progress and we blocked, retry allocation now that
  //memory has been freed
  mirror::Object* ptr = TryToAllocate<true, false>(self, allocator,
      alloc_size, Bytes_allocated, usable_size,
      Bytes_tl_bulk_allocated);
  if (ptr != nullptr) {
    return ptr;
  }
}

collector::GcType tried_type = next_gc_type_;
const bool gc_ran =
    CollectGarbageInternal(tried_type, kGcCauseForAlloc,
                           false) != collector::kGcTypeNone;
if ((was_default_allocator && allocator != GetCurrentAllocator()) ||
    (!instrumented && EntrypointsInstrumented())) {
  return nullptr;
}
if (gc_ran) {
  mirror::Object* ptr = TryToAllocate<true, false>(self, allocator,
   alloc_size, Bytes_allocated, usable_size, Bytes_tl_bulk_allocated);
  if (ptr != nullptr) {
    return ptr;
  }
}

//Loop through our different Gc types and try to Gc until we get
//enough free memory
for (collector::GcType gc_type : gc_plan_) {
  if (gc_type == tried_type) {
    continue;
  }
```

```cpp
    //Attempt to run the collector, if we succeed, re-try the
    //allocation
    const bool plan_gc_ran =
        CollectGarbageInternal(gc_type, kGcCauseForAlloc, false) != collector::kGcTypeNone;
    if ((was_default_allocator && allocator != GetCurrentAllocator()) ||
        (!instrumented && EntrypointsInstrumented())) {
      return nullptr;
    }
    if (plan_gc_ran) {
      //Did we free sufficient memory for the allocation to succeed
      mirror::Object* ptr = TryToAllocate<true, false>(self, allocator, alloc_size, Bytes
_allocated, usable_size, Bytes_tl_bulk_allocated);
      if (ptr != nullptr) {
        return ptr;
      }
    }
  }
  //Allocations have failed after GCs; this is an exceptional state.
  //Try harder, growing the heap if necessary
  mirror::Object* ptr = TryToAllocate<true, true>(self, allocator,
      alloc_size, Bytes_allocated, usable_size, Bytes_tl_bulk_allocated);
  if (ptr != nullptr) {
    return ptr;
  }
  //Most allocations should have succeeded by now, so the heap is really
  //full, really fragmented, or the requested size is really big. Do
  //another GC, collecting SoftReferences this time. The
  //VM spec requires that all SoftReferences have been collected and
  //cleared before throwing OOME
  VLOG(gc) << "Forcing collection of SoftReferences for "
           << PrettySize(alloc_size)
           << " allocation";
  //TODO: Run finalization, but this may cause more allocations to
  //occur. We don't need a WaitForGcToComplete here either
  DCHECK(!gc_plan_.empty());
  CollectGarbageInternal(gc_plan_.back(), kGcCauseForAlloc, true);
  if ((was_default_allocator && allocator != GetCurrentAllocator()) ||
      (!instrumented && EntrypointsInstrumented())) {
    return nullptr;
  }
  ptr = TryToAllocate<true, true>(self, allocator, alloc_size,
      Bytes_allocated, usable_size, Bytes_tl_bulk_allocated);
  if (ptr == nullptr) {
    const uint64_t current_time = NanoTime();
    switch (allocator) {
      case kAllocatorTypeRosAlloc:
```

```cpp
    //Fall-through
  case kAllocatorTypeDlMalloc: {
    if (use_homogeneous_space_compaction_for_oom_ &&
        current_time - last_time_homogeneous_space_compaction_by_oom_ >
        min_interval_homogeneous_space_compaction_by_oom_) {
      last_time_homogeneous_space_compaction_by_oom_ = current_time;
      HomogeneousSpaceCompactResult result =
          PerformHomogeneousSpaceCompact();
      //Thread suspension could have occurred
      if ((was_default_allocator && allocator !=
          GetCurrentAllocator()) ||
          (!instrumented && EntrypointsInstrumented())) {
        return nullptr;
      }
      switch (result) {
        case HomogeneousSpaceCompactResult::kSuccess:
          //If the allocation succeeded, we delayed an oom
          ptr = TryToAllocate<true, true>(self, allocator, alloc_size,
              Bytes_allocated, usable_size, Bytes_tl_bulk_allocated);
          if (ptr != nullptr) {
            count_delayed_oom_++;
          }
          break;
        case HomogeneousSpaceCompactResult::kErrorReject:
          //Reject due to disabled moving GC
          break;
        case HomogeneousSpaceCompactResult::kErrorVMShuttingDown:
          //Throw OOM by default
          break;
        default: {
          UNIMPLEMENTED(FATAL)
              << "homogeneous space compaction result: "
              << static_cast<size_t>(result);
          UNREACHABLE();
        }
      }
    }
    //Always print that we ran homogeneous space compation since
    //this can cause jank
    VLOG(heap) << "Ran heap homogeneous space compaction, "
        << " requested defragmentation "
        << count_requested_homogeneous_space_compaction_.load()
        << " performed defragmentation "
        << count_performed_homogeneous_space_compaction_.load()
        << " ignored homogeneous space compaction "
        << count_ignored_homogeneous_space_compaction_.load()
        << " delayed count = "
```

```cpp
                          << count_delayed_oom_.load();
      }
      break;
    }
    case kAllocatorTypeNonMoving: {
      if (kUseReadBarrier) {
        //DisableMovingGc() isn't compatible with CC
        break;
      }
      //Try to transition the heap if the allocation failure was due to
      //the space being full
      if (!IsOutOfMemoryOnAllocation(allocator, alloc_size, false)) {
        //If we aren't out of memory then the OOM was probably from the
        //non moving space being full. Attempt to disable compaction
        //and turn the main space into a non moving space
        DisableMovingGc();
        //Thread suspension could have occurred
        if ((was_default_allocator && allocator !=
            GetCurrentAllocator()) ||
            (!instrumented && EntrypointsInstrumented())) {
          return nullptr;
        }
        //If we are still a moving GC then something must have caused
        //the transition to fail
        if (IsMovingGc(collector_type_)) {
          MutexLock mu(self, *gc_complete_lock_);
          //If we couldn't disable moving GC, just throw OOME and return
          //null
          LOG(WARNING) << "Couldn't disable moving GC with disable GC
              count " << disable_moving_gc_count_;
        } else {
          LOG(WARNING) << "Disabled moving GC due to the non moving space
              being full";
          ptr = TryToAllocate<true, true>(self, allocator, alloc_size,
              Bytes_allocated, usable_size, Bytes_tl_bulk_allocated);
        }
      }
      break;
    }
    default: {
      //Do nothing for others allocators
    }
  }
}
//If the allocation hasn't succeeded by this point, throw an OOM
//error
```

```
    if (ptr == nullptr) {
        ThrowOutOfMemoryError(self, alloc_size, allocator);
    }
    return ptr;
}
```

CollectGarbageInternal()函数比较重要,在后续的GC触发中也会用到,有关这个函数的实现在后续进行分析。

2. kGcCauseBackground

kGcCauseBackground是触发后台执行的GC,为了确保每次分配空间都有空间可用,可以将其理解为并发GC,也就是说回收器和分配器同时在执行。这个触发动作是在分配内存成功之后进行的触发。在实际代码中是通过AllocObjectWithAllocator中执行的CheckConcurrentGCForJava()函数实现相关动作的,CheckConcurrentGCForJava()函数的实现位于art/Runtime/gc/heap-inl.h中,代码如下:

```
//第7章/heap-inl.h
inline void Heap::CheckConcurrentGCForJava(Thread* self,
    size_t new_num_Bytes_allocated, ObjPtr<mirror::Object>* obj) {
    if (UNLIKELY(ShouldConcurrentGCForJava(new_num_Bytes_allocated))) {
        RequestConcurrentGCAndSaveObject(self, false /* force_full */, obj);
    }
}
```

这里通过ShouldConcurrentGCForJava()函数的条件判断调用了RequestConcurrentGCAndSaveObject()函数。ShouldConcurrentGCForJava()函数用于判断是否进行并发GC的请求。RequestConcurrentGCAndSaveObject()函数则直接进行并发GC请求,它的实现位于art/Runtime/gc/heap.cc中,代码如下:

```
//第7章/heap.cc
void Heap::RequestConcurrentGCAndSaveObject(Thread* self,
    bool force_full, ObjPtr<mirror::Object>* obj) {
    StackHandleScope<1> hs(self);
    HandleWrapperObjPtr<mirror::Object> wrapper(hs.NewHandleWrapper(obj));
    RequestConcurrentGC(self, kGcCauseBackground, force_full);
}
```

在上述代码中,通过调用同处于heap.cc文件中的RequestConcurrentGC()函数,请求进行并发GC。RequestConcurrentGC()函数的代码如下:

```
//第7章/heap.cc
void Heap::RequestConcurrentGC(Thread* self, GcCause cause,
                                bool force_full) {
```

```
    if (CanAddHeapTask(self) &&
        concurrent_gc_pending_.CompareAndSetStrongSequentiallyConsistent(
        false, true)) {
      task_processor_ -> AddTask(self, new ConcurrentGCTask(NanoTime(),
                                                            cause,
                                                            force_full));
    }
  }
```

在 RequestConcurrentGC() 函数中,新建了一个 ConcurrentGCTask 对象,并且将其作为一个任务添加到了 task_processor_ 中。task_processor_ 是 TaskProcessor 类型的智能指针。TaskProcessor 类的实现在 art/Runtime/gc/task_processor.h 中,它有一个 RunAllTasks() 函数,会将 task 取出并执行其 Run() 函数。ConcurrentGCTask 类的 Run() 函数是这个任务的主要动作,它也位于 heap.cc 中,代码如下:

```
//第 7 章/heap.cc
class Heap::ConcurrentGCTask : public HeapTask {
public:
  ConcurrentGCTask(uint64_t target_time, GcCause cause, bool force_full)
      : HeapTask(target_time), cause_(cause), force_full_(force_full) {}
  void Run(Thread* self) override {
    gc::Heap* heap = Runtime::Current()->GetHeap();
    heap->ConcurrentGC(self, cause_, force_full_);
    heap->ClearConcurrentGCRequest();
  }
}
```

这里面的核心环节是 Run() 函数调用了 Heap 类的 ConcurrentGC() 函数,它也位于 heap.cc 中,代码如下:

```
//第 7 章/heap.cc
void Heap::ConcurrentGC(Thread* self, GcCause cause, bool force_full) {
  if (!Runtime::Current()->IsShuttingDown(self)) {
    //Wait for any GCs currently running to finish
    if (WaitForGcToComplete(cause, self) == collector::kGcTypeNone) {
      //If we can't run the GC type we wanted to run, find the next
      //appropriate one and try that instead. E.g. can't do partial, so
      //do full instead
      collector::GcType next_gc_type = next_gc_type_;
      //If forcing full and next gc type is sticky, override with a non-
      //sticky type
      if (force_full && next_gc_type == collector::kGcTypeSticky) {
        next_gc_type = NonStickyGcType();
      }
```

```
        if (CollectGarbageInternal(next_gc_type, cause, false) ==
          collector::kGcTypeNone) {
          for (collector::GcType gc_type : gc_plan_) {
            //Attempt to run the collector, if we succeed, we are done
            if (gc_type > next_gc_type &&
                CollectGarbageInternal(gc_type, cause, false) != collector::kGcTypeNone) {
              break;
            }
          }
        }
      }
    }
  }
}
```

从上述代码可以看到,这个函数里面最终执行了 GC 动作,这也是通过调用 CollectGarbageInternal() 函数进行具体操作的。

3. kGcCauseExplicit

kGcCauseExplicit 采用显式方式触发 GC,主要通过函数 System.gc() 接口显式地调用,然后触发 GC。在实际代码中,该过程的动作主要通过 CollectGarbage() 函数进行实现的,该函数的声明位于 art/Runtime/gc/heap.h 中,代码如下:

```
//第 7 章/heap.h
//Initiates an explicit garbage collection
  void CollectGarbage(bool clear_soft_references,
                    GcCause cause = kGcCauseExplicit)
      REQUIRES(!*gc_complete_lock_, !*pending_task_lock_);
```

可以看到,CollectGarbage() 函数的第二个参数缺省值为 kGcCauseExplicit,已经将其定义为显式触发。CollectGarbage() 函数的具体实现位于 art/Runtime/gc/heap.cc 中,代码如下:

```
//第 7 章/heap.cc
void Heap::CollectGarbage(bool clear_soft_references, GcCause cause) {
  //Even if we waited for a GC we still need to do another GC since
  //weaks allocated during the last GC will not have necessarily been
  //cleared
  CollectGarbageInternal(gc_plan_.back(), cause, clear_soft_references);
}
```

在 CollectGarbage() 函数中也通过调用 CollectGarbageInternal() 函数来实现垃圾回收。这和上面两个 GC 触发一样,也就是说在这里分析的 3 个 GC 触发最终都会通过调用 CollectGarbageInternal() 函数进行 GC。

CollectGarbageInternal() 函数的实现位于 art/Runtime/gc/heap.cc 中,它通过判断相

关条件,为回收器的运行准备了所需要的条件,最终调用回收器进行内存回收,代码如下:

```cpp
//第 7 章/heap.cc
collector::GcType Heap::CollectGarbageInternal(collector::GcType gc_type,
    GcCause gc_cause, bool clear_soft_references) {
  Thread* self = Thread::Current();
  Runtime* Runtime = Runtime::Current();
  //If the heap can't run the GC, silently fail and return that no GC
  //was run
  switch (gc_type) {
    case collector::kGcTypePartial: {
      if (!HasZygoteSpace()) {
        return collector::kGcTypeNone;
      }
      break;
    }
    default: {
      //Other GC types don't have any special cases which makes them not
      //runnable. The main case here is full GC
    }
  }
  ScopedThreadStateChange tsc(self, kWaitingPerformingGc);
  //TODO: Clang prebuilt for r316199 produces bogus thread safety
  //analysis warning for holding both exclusive and shared lock in the
  //same scope. Remove the assertion as a temporary workaround
  //http://b/71769596
  //Locks::mutator_lock_->AssertNotHeld(self)
  if (self->IsHandlingStackOverflow()) {
    //If we are throwing a stack overflow error we probably don't have
    //enough remaining stack space to run the GC
    return collector::kGcTypeNone;
  }
  bool compacting_gc;
  {
    gc_complete_lock_->AssertNotHeld(self);
    ScopedThreadStateChange tsc2(self, kWaitingForGcToComplete);
    MutexLock mu(self, *gc_complete_lock_);
    //Ensure there is only one GC at a time
    WaitForGcToCompleteLocked(gc_cause, self);
    compacting_gc = IsMovingGc(collector_type_);
    //GC can be disabled if someone has a used GetPrimitiveArrayCritical
    if (compacting_gc && disable_moving_gc_count_ != 0) {
      LOG(WARNING) << "Skipping GC due to disable moving GC count "
                   << disable_moving_gc_count_;
      return collector::kGcTypeNone;
    }
```

```cpp
    if (gc_disabled_for_shutdown_) {
      return collector::kGcTypeNone;
    }
    collector_type_running_ = collector_type_;
  }
  if (gc_cause == kGcCauseForAlloc && Runtime->HasStatsEnabled()) {
    ++Runtime->GetStats()->gc_for_alloc_count;
   ++self->GetStats()->gc_for_alloc_count;
  }
  const size_t Bytes_allocated_before_gc = GetBytesAllocated();

  DCHECK_LT(gc_type, collector::kGcTypeMax);
  DCHECK_NE(gc_type, collector::kGcTypeNone);

  collector::GarbageCollector * collector = nullptr;
  //TODO: Clean this up
  if (compacting_gc) {
    DCHECK(current_allocator_ == kAllocatorTypeBumpPointer ||
           current_allocator_ == kAllocatorTypeTLAB ||
           current_allocator_ == kAllocatorTypeRegion ||
           current_allocator_ == kAllocatorTypeRegionTLAB);
    switch (collector_type_) {
      case kCollectorTypeSS:
        //Fall-through
      case kCollectorTypeGSS:
        semi_space_collector_->SetFromSpace(bump_pointer_space_);
        semi_space_collector_->SetToSpace(temp_space_);
        semi_space_collector_->SetSwapSemiSpaces(true);
        collector = semi_space_collector_;
        break;
      case kCollectorTypeCC:
        if (use_generational_cc_) {
          //TODO: Other threads must do the flip checkpoint before they
          //start poking at active_concurrent_copying_collector_. So we
          //should not concurrency here
          active_concurrent_copying_collector_ = (
              gc_type == collector::kGcTypeSticky) ?
              young_concurrent_copying_collector_ : concurrent_copying_collector_;
          DCHECK(active_concurrent_copying_collector_->RegionSpace() == region_space_);
        }
        collector = active_concurrent_copying_collector_;
        break;
      default:
        LOG(FATAL) << "Invalid collector type "
                   << static_cast<size_t>(collector_type_);
  }
```

```cpp
    if (collector != active_concurrent_copying_collector_) {
      temp_space_->GetMemMap()->Protect(PROT_READ | PROT_WRITE);
      if (kIsDebugBuild) {
        //Try to read each page of the memory map in case mprotect didn't
        //work properly b/19894268
        temp_space_->GetMemMap()->TryReadable();
      }
      CHECK(temp_space_->IsEmpty());
    }
    gc_type = collector::kGcTypeFull; //TODO: Not hard code this in
  } else if (current_allocator_ == kAllocatorTypeRosAlloc ||
      current_allocator_ == kAllocatorTypeDlMalloc) {
    collector = FindCollectorByGcType(gc_type);
  } else {
    LOG(FATAL) << "Invalid current allocator " << current_allocator_;
  }

  CHECK(collector != nullptr)
      << "Could not find garbage collector with collector_type = "
      << static_cast<size_t>(collector_type_) << " and gc_type = "
      << gc_type;
  collector->Run(gc_cause, clear_soft_references || Runtime->IsZygote());
  total_objects_freed_ever_ += GetCurrentGcIteration()->GetFreedObjects();
  total_Bytes_freed_ever_ += GetCurrentGcIteration()->GetFreedBytes();
  RequestTrim(self);
  //Collect cleared references
  SelfDeletingTask* clear = reference_processor_->CollectClearedReferences(self);
  //Grow the heap so that we know when to perform the next GC
  GrowForUtilization(collector, Bytes_allocated_before_gc);
  LogGC(gc_cause, collector);
  FinishGC(self, gc_type);
  //Actually enqueue all cleared references. Do this after the GC has
  //officially finished since otherwise we can deadlock
  clear->Run(self);
  clear->Finalize();
  //Inform DDMS that a GC completed
  Dbg::GcDidFinish();

  old_native_Bytes_allocated_.store(GetNativeBytes());

  //Unload native libraries for class unloading. We do this after
  //calling FinishGC to prevent deadlocks in case the JNI_OnUnload
  //function does allocations
  {
    ScopedObjectAccess soa(self);
    soa.Vm()->UnloadNativeLibraries();
```

```
    }
    return gc_type;
}
```

这里的核心路径通过 collector::GarbageCollector * collector = nullptr;声明了一个collector,并且根据参数和各方面的条件,为其赋值目前要使用的回收器,然后执行collector的 Run()函数进行垃圾回收。前面在介绍回收器的实现的时候介绍过,回收器的 Run()函数是其核心函数,其所执行的内容都在这里面,所以到这一步,也就追踪到了 GC 回收器的真正调用,基本上梳理清了回收器的触发流程。

前文介绍了分配器和回收的设计与实现,本部分分别对分配器和回收器的使用进行了分析,理清了分配器和回收器的调用接口和基本流程,方便读者理解在实际过程中如何使用分配器和回收器。

7.6 小结

本章主要对 ART 的 GC 进行了分析,GC 是虚拟机中很重要的一个组成部分。本章从 GC 的基本内容入手,介绍了 ART GC 的几种基本方案,然后对 ART GC 中分配器和回收器的实现进行了分析,最后通过分析 ART GC 分配器和回收器的使用,理清了实际调用分配器和回收器的基本流程。

此外,由于篇幅限制,本章并未过多地介绍与 GC 理论相关的内容,读者如果有需要可以查阅笔者所推荐的相关资料。对于 ART GC 中的很多细节内容,本章也并未进行介绍,而是将焦点集中在了框架和流程上,读者可以根据自己的需要,在本章内容的基础之上进一步深入分析源码。

参 考 文 献

［1］ 邓凡平.深入理解 Android-Java 虚拟机 ART[M].北京：机械工业出版社,2019.
［2］ 钟世礼.深入解析 Android 虚拟机[M].北京：人民邮电出版社,2016.
［3］ 李晓峰.虚拟机设计与实现[M].单业,译.北京：人民邮电出版社,2020.
［4］ JONES R,HOSKING A,MOSS E.垃圾回收算法手册——自动内存管理的艺术[M].王雅光,薛迪,译.北京：机械工业出版社,2016.
［5］ 中村成洋,相川光.垃圾回收的算法与实现[M].丁灵,译.北京：人民邮电出版社,2016.
［6］ 中村成洋.深入 Java 虚拟机[M].吴炎昌,杨文轩,译.北京：人民邮电出版社,2021.
［7］ COOPER K D,TORCZON L.编译器设计[M].郭旭,译.2 版.北京：人民邮电出版社,2013.
［8］ MUCHNICK S S. Advanced Compiler Design and Implementation(影印版)[M].北京：机械工业出版社,2003.
［9］ 调试 ART 垃圾回收[J/OL]. https://source.android.com/devices/tech/dalvik/gc-debug.
［10］ Dalvik(software)[J/OL]. https://zh.wikipedia.org/wiki/Dalvik.
［11］ Android Runtime[J/OL]. https://zh.wikipedia.org/wiki/Android_Runtime.
［12］ Dominator(graph_theory)[J/OL]. https://en.wikipedia.org/wiki/Dominator_(graph_theory).
［13］ Static single assignment form[J/OL]. https://en.wikipedia.org/wiki/Static_single_assignment_form.
［14］ Register allocation[J/OL]. https://en.wikipedia.org/wiki/Register_allocation.

后　　记

　　2020年，为了准备一个项目，笔者花费了很多时间在阅读ART的源码，在阅读的过程中整理和记录了不少笔记。在这些笔记的基础上，不断地进行添加和修改，使其内容逐渐体系化，最终有了这本书的骨架。2021年笔者陆续又修改了一段时间，终于完成了本书的内容。

　　目前市面上有关Android源码分析或者ART源码分析的书并不少，其中所涉及的ART版本都较旧，所面向对象多为App开发者。本书希望能面向更广泛的读者群体，给读者不一样的体验，让读者更好地理解ART中的编译和运行时方面的内容。

　　Android是一个庞大的系统，ART又是其中的核心，在本书的写作过程中不可避免地遇到了很多问题。很多朋友为笔者提供了帮助，在此专门表示感谢，由于各种因素就不一一具名了。虽然尽力想把本书写得尽量完善一些，但是肯定还存在诸多不足，还望大家海涵。

　　目前，技术书籍的写作并非一件易事。不仅需要投入大量的时间和精力，还需要各方面的支持和配合。在此，专门感谢中国科学院软件研究所智能软件中心及PLCT实验室各位领导及同仁的大力支持，为本书的编写提供了各方面的便利条件，最终使本书能够有机会出版。

　　赵佳霓女士作为本书的责任编辑，从立项直至出版为本书付出了很多，在此专门致谢。

图 书 推 荐

书　名	作　者
鸿蒙应用程序开发	董昱
鸿蒙操作系统开发入门经典	徐礼文
鸿蒙操作系统应用开发实践	陈美汝、郑森文、武延军、吴敬征
华为方舟编译器之美——基于开源代码的架构分析与实现	史宁宁
鲲鹏架构入门与实战	张磊
华为 HCIA 路由与交换技术实战	江礼教
Flutter 组件精讲与实战	赵龙
Flutter 组件详解与实战	［加］王浩然（Bradley Wang）
Flutter 实战指南	李楠
Dart 语言实战——基于 Flutter 框架的程序开发（第 2 版）	亢少军
Dart 语言实战——基于 Angular 框架的 Web 开发	刘仕文
IntelliJ IDEA 软件开发与应用	乔国辉
Vue＋Spring Boot 前后端分离开发实战	贾志杰
Vue.js 企业开发实战	千锋教育高教产品研发部
Python 人工智能——原理、实践及应用	杨博雄主编，于营、肖衡、潘玉霞、高华玲、梁志勇副主编
Python 深度学习	王志立
Python 异步编程实战——基于 AIO 的全栈开发技术	陈少佳
Python 数据分析从 0 到 1	邓立文、俞心宇、牛瑶
物联网——嵌入式开发实战	连志安
智慧建造——物联网在建筑设计与管理中的实践	［美］周晨光（Timothy Chou）著；段晨东、柯吉译
TensorFlow 计算机视觉原理与实战	欧阳鹏程、任浩然
分布式机器学习实战	陈敬雷
计算机视觉——基于 OpenCV 与 TensorFlow 的深度学习方法	余海林、翟中华
深度学习——理论、方法与 PyTorch 实践	翟中华、孟翔宇
深度学习原理与 PyTorch 实战	张伟振
ARKit 原生开发入门精粹——RealityKit＋Swift＋SwiftUI	汪祥春
HoloLens 2 开发入门精要——基于 Unity 和 MRTK	汪祥春
Altium Designer 20 PCB 设计实战（视频微课版）	白军杰
Cadence 高速 PCB 设计——基于手机高阶板的案例分析与实现	李卫国、张彬、林超文
Octave 程序设计	于红博
SolidWorks 2020 快速入门与深入实战	邵为龙
SolidWorks 2021 快速入门与深入实战	邵为龙
UG NX 1926 快速入门与深入实战	邵为龙
西门子 S7-200 SMART PLC 编程及应用（视频微课版）	徐宁、赵丽君
三菱 FX3U PLC 编程及应用（视频微课版）	吴文灵
全栈 UI 自动化测试实战	胡胜强、单镜石、李睿
pytest 框架与自动化测试应用	房荔枝、梁丽丽
软件测试与面试通识	于晶、张丹
深入理解微电子电路设计——电子元器件原理及应用（原书第 5 版）	［美］理查德・C. 耶格（Richard C. Jaeger）、［美］特拉维斯・N. 布莱洛克（Travis N. Blalock）著；宋廷强译
深入理解微电子电路设计——数字电子技术及应用（原书第 5 版）	［美］理查德・C. 耶格（Richard C. Jaeger）、［美］特拉维斯・N. 布莱洛克（Travis N. Blalock）著；宋廷强译
深入理解微电子电路设计——模拟电子技术及应用（原书第 5 版）	［美］理查德・C. 耶格（Richard C. Jaeger）、［美］特拉维斯・N. 布莱洛克（Travis N. Blalock）著；宋廷强译